Tarraforming Earth

Are They Back to Stay?

Jessie R. Dubose

iUniverse, Inc.
Bloomington

Tarraforming Earth
Are They Back to Stay?

iUniverse books may be ordered through booksellers or by contacting:

iUniverse
1663 Liberty Drive
Bloomington, IN 47403
www.iuniverse.com
1-800-Authors (1-800-288-4677)

ISBN: 978-1-4697-5299-0 (sc)
ISBN: 978-1-4697-5300-3 (hc)
ISBN: 978-1-4697-5301-0 (e)

Library of Congress Control Number: 2012901563

Printed in the United States of America

iUniverse rev. date: 1/24/2012

Contents

ABOUT AUTHOR

THIS AUTHOR IS NEITHER a scientist nor religious. He has a BSEE from California State College and has worked his entire career as an electrical engineer and software engineer in Dallas, Texas.

He is proposing a scientific theory for which he has no scientific credentials. Since those with credentials have not done so, I feel that it is my responsibility to mankind to do so.

He has been studying unknown phenomena in the public media for the past forty years. He has decided that it is about time that all the How,Who,When,What,Why questions should be taken as all being related. All these questions can be answered by a cohesive theory.

His TARRAFORMERING EARTH theory will allow a framework to accept the HWWWW questions. Many prominent scientists will support this theory because it will shield them from reprisals from the government and their peers.

It is his belief that the publishers of videos and authors of books about ancient aliens, ancient astronauts, Bigfoot and UFOs can at last have a common explanation. The UFOs are here and now and can not be explained away.

It is accepted by many that an exterrestrial civilization created man. The author believes that it created all life on earth since its beginning of habitability. This book is not intended to be a reference for research. He intends to draw a time line for earth's evolution from the beginning to present day. .

Although the exterrestrial civilization has not had any direct contact with man in the last 2000 years, it seems to have decided to continue monitoring man's evolution since WWII.

Anyone wishing to support or reject the theory can reach the author at jessiedubose@yahoo.com.

FORWARD

THE WORD TERRAFORMING WAS coined by science to convey the process of converting alien environments to mimic that of earth. This presentation changes the spelling to "tarraforming" to indicate the inverse of the process. This presentation suggests that the effort was not to mimic the home planet but to install life forms that could exist on an evolving planet For this scenario to be plausible there must be an extraterrestrial civilization (XTC) available to perform the task. The members of which will be referred to as the XTB. The XTBs are what religion refers to as pagan gods and science refers to as mythical gods. Regardless of the definition they figure prominently in ancient history.

The popular media is replete with "How,Who,When,What" presentations for the many unexplained phenomena. Since the presentations are not directly related to other presentations it is easy to debunk the presentation. The tarraforming theory will attempt to allow all the disparate presentations to be tied into a comprehensive whole.

The XTC was involved in a tarraforming project millions of years before the advent of man. It will be purposed that the tarraforming project was responsible for restoring life forms following catastrophes. It could be assumed that the XTC was aware of the process of evolution long before Darwin proposed his theory. It is assumed that the tarraformers would install a life form and monitored its evolution. The observation periods could be as long as millions of years. The evolution process is frequently interrupted by various catastrophes. This is when they wait for the dust to settle before installing new life forms suited to the new environment.

Some 450 thousand years ago; a date established by Zecharia Sitchin in his numerous books, the XTC inserted a mining operation into the tarraforming operation. This was probably seen as an unnatural catastrophe by the tarraformers. The miners built the many megalithic structures and cities. The tarraformers had no need for cities. The mining operation resulted in the creation of a mine worker from an existing life form which was not an ape. The increased intelligence required of the worker's evolution be advanced by millions of years. This mine worker is likely what is known as Neanderthal. Eventually the mining operation was terminated. This is probably the reason for the abandonment of the many ancient cities and why no new construction occurred.

The tarraformers commenced to clean up the mess made by the miners. They had to decide how to fit the mine workers into their new project. It was decided that his intelligence had not been advanced enough to exist without intervention. It was then decided to advance his intelligence even further. This resulted in what is known as the creation of man.

The second phase of man's advancement probably resulted in what the bible refers to as the Garden of Eden. The Garden of Eden can be thought of as a laboratory similar to a game preserve. Once a life form had sufficiently developed it was transferred to the local environment. The religious reason given for the transfer was disobedience to the responsible XTB (God).

When the tarraformers finally got their project back on track all direct contact with man was terminated. This required that most XTBs be transferred back to the home planet. This was the beginning of monotheism and the end of paganism as well as the final abandonment of ancient cities. This required that religion advance the belief in an external unseen God that no longer perform miracles. After a period of years it was recognized that it was difficult to convince the population to believe in an external unseen god.

An attempt to bolster the belief in an external god was implemented. This involved the creation of a demigod purported to be the son of God who would remain on earth. When the priest and roman emperor saw Jesus as a threat they crucified him. This is one of the most significant points in man's evolution. Although Jesus did not survive, the belief in an external God did because of him. Religion guided the evolution of man for thousand of years after the Jesus event. It is likely that some religious priest had indirect contact with XTBs(angels) in following years.

The second leap in man's evolution happened about 500 years ago with the advent of what is known as the scientific method. This is when moral teachings of religion lost to the philosophy of materialism. Religion and science have been at odds ever since. This has led to the chauvinistic view that man's evolution was not effected by a higher intelligence. Religion says believe this because we say it is true while scientist say believe this because we have a theory. Free thinkers say I don't believe either of you because your proposals do not make sense.

The latest leap in man's evolution is happening now. This is evidenced by UFO reports that science cannot explain and stoops so low as to ridicule. The purpose of the ridicule is to indoctrinate the population into believing that the XTC does not exist.

A passage from the bible supports the possibility of extraterrestrial life. This can be found in John 14:2-3 as repeated here.

In my father's house are many mansions: if it were not so, I would have told you. I go to prepare a place for you. And if I go and prepare a place for you, I will come again, and receive you unto myself; that where I am, there ye may be also.

The following lists the research sources found most helpful.

1. Erich Von Daniken books.
2. Zecharia Sitchin books.
3. Ancient Aliens video.
4. UFO Hunters video.
5. Hundreds of others

If you accept the theory and would like to contribute to the next book, contact the author at jessiedubose@yahoo.com. Your contribution will be fully noted if included and compensation will be appropriate.

Section-1
EARTH'S BEGINNING

BILLIONS OF YEARS AGO there existed an extraterrestrial civilization somewhere in the universe. Henceforth, this civilization will be referred to as the XTC. Members of this civilization will be referred to as extraterrestrial beings (XTB). The XTB are referred to as pagan gods by religion and mythical gods by science. In reality there may be more than one advanced civilization. Our XTC observed the formation of many solar systems. One such solar system was that of earth.

Much of the knowledge of the early solar system was documented by the Sumerian civilization who in turn acquired their knowledge from the XTC. Modern day scientist spend most of their time trying to ignore the information or trying to debunk it. It is ironical that many of modern cosmological theories are very similar to the Sumerian.

Chapter one of genesis in the holy bible explains the time line for the creation of the solar system and not the universe as some think. The time line is laid out in days. This is the most used point of argument for the debunkers. It must be realized that when the knowledge was transmitted to man one of the measurements of time understood by most was a day. Early man did not have any concept of billions and millions. From this it may be assumed that a day was used to explain time segments of events which could have been billions of years.

The second verse of genesis implies that the planets were present before the ignition of the sun as stated in verse three. Verses four through ten deal with the geological processes that affected the early earth. Verse eleven can be seen as the beginning of the tarraforming project's installation of life forms. This assertion presents the possibility that there are many solar systems in the process of forming but cannot be detected because their sun has not ignited or inadequacy of detection equipment. Regardless how you slice it, the time line laid out in genesis is very similar to that of modern cosmology.

2-Birth Of Earth

The early solar system was a very chaotic place. There was a watery planet in the solar system plane that shared its orbit with a planet that had an orbit that was inclined to the plane of the solar system. This planet was referred to as the planet of crossing (POC). Every 3600 years the POC passes through the solar system plane.

Billions of years ago the POC occupied a point in its orbit that was coincident with a point in the orbit of the watery planet. The POC had numerous satellites that rotated opposite those of in plane planets. It was one of these satellites that collided with the watery planet. The remnants of the watery planet along with its moon were transferred to a new orbit, and became known as earth. This collision could also have changed the orbit of the POC. This solves the scientist's question of, "how did the earth get it's water?". That is to say that the earth had its water from the beginning and did not acquire its water by comet impacts. Robotic probes have verified that water rich moons exist throughout the solar system, so why could have a water rich planet not have existed.

The shared orbit of the two planets between Mars and Jupiter contain debris from the collision and became known as the hammered bracelet by the Sumerians and the asteroid belt by modern astronomers. An early mathematician named Bode developed a planetary law that predicted the orbits of planets. It so happens that the law predicts that a planet should exist in the asteroid belt. Modern astronomers purpose that the asteroid belt is building material for a planet that did not form. This building material is theorized as primordial gases clumping together. It is known that if gases clump together the process can not produce iron and nickel, of which many asteroids are composed. The theory also includes the possibility of asteroid collisions. One has to ask himself, how does a collision occur if every thing is moving in the same direction and at the same speed? The answer may be a periodic gravitational force exerted by the POC.

The POC continues to pass through the hammered bracelet at which time many of the asteroids are ejected into different orbits. Some of these orbits intersected the new orbit of earth. This would explain the many asteroid impacts experienced by the early earth and its moon. If one of these asteroids was of sufficient size, it could have caused extinctions. If the earth is sufficiently close to the POC when it passes through the solar plane, extreme tidal forces could result.

Some researchers purpose that XTBs mentioned in the ancient text were inhabitants of the POC. This is not a rational assumption since the POC spends much of its orbital cycle in the far reaches of the solar system where sunlight is negligible and the temperature is very low. A more reasonable assumption is that the XTBs were inhabitants of a stable planet around another star. The XTBs were most active on earth when the POC entered in close proximately to earth, so they could monitor the ensuing events. One such event was probably the worldwide flood documented in the bible.

The worldwide flood mentioned in many religious texts could have been the result of the POC passing close enough to earth, so as to create massive tides. Many of the minor catastrophes may also have been caused by the passage of the POC. The severity of the catastrophe depends on the distance between the POC and earth at the time of crossing. The biblical account of the flood speaks of an extended period of heavy rain as the cause. Heavy rain was observed because of association with high tides caused by the gravitational effect of the POC. The high tides caused the flooding, not the rain.

If rain had caused the flood, one must ask the question where did the immense amounts of water come from and where did it go?

The ancient text of the Indus Valley have been relegated to religious myth by science. The most studied of these are the Mahabharata and Ramayana. The reason it was easy for science to relegate the text to myth was the fact that they were composed only in 200 BC from oral traditions. A supporting argument was that events described were impossible. It later turned out some of the events were shown to be possible with the development of nuclear weapons at the end of WWII. Our feeble attempts at space travel have lead the uninformed to state that if we can not do it, no one else can either.

The events in the text that is the hardest to accept is the description of areal battles in those ancient times. If one eliminates the possibility that the composers read science fiction novels, then the only conclusion is that another XTC tried to wrest control of earth from the resident XTC. It is not realistic to assume the battles were between adversarial resident XTBs because of the scope of the battles.

Much time has been expended on analysis of ancient text to glean ancient knowledge. Just as much time has been expended in an effort to disparage ancient knowledge. With our superior knowledge achieved through pot shards and building foundations we know what the ancient text really tried to convey. Ancient scribes did not have the vocabulary or knowledge to describe events witnessed, therefore it must be the one thing that everyone could identify with, that is the gods. The gods are the debunker's main point of attack because of the religion/science conflict.

Many scientific disciplines have evolved to advance the study of the earth's early history. These studies revolve around what is known as fossil evidence. Fossil evidence is anything that leaves evidence of its prior existence, not just bones. Scientist analyze the evidence and develop theories to explain it. Some of these theories make sense, while others are unrealistic. A problem arises when a previous unsound theory is used to explain new evidence to develop a new theory.

3-EVOLUITION

ANCIENT TEXT COVERS A very small time span in the earths evolution. However, they are the only markers of intelligent observations of the past. If this is not available the only alternative is theories. Remember that theories are guesses. Some of these theories can be accepted as fact depending on the notoriety of the sponsor. Some of these theories will be questioned in the following segments.

This is where the tarraformers enter the discussion. The tarraforming theory proposes that the earth's ecosystem has been manipulated for millions of years and maybe as many as a billion. The tarraformers mode of operation is to install a life form and monitor its evolution. This may require thousand of years. An earthly scientist recognized this process and developed his famous and accepted theory.

In the early 1800s Charles Darwin developed the theory that all species of plants and animals developed from earlier forms by hereditary transmission of slight variations to successive generations. The forms that survive are those that are best adapted to the environment. The theory has been proven to explain the variations within a species.

It is theorized that all life on earth evolved from a single cell organism which in turn evolved from chemical elements or spores from asteroids. Sloshing of tides were also required to make the theory work. It would be logical to assume that plants evolved before animals since animals would require a food source. Very little research has been published about plants evolution during this time. This requires two unrelated life forms to develop independently from single cell organisms.

This problem can be solved by assuming that the XTC would install plant lifeforms and wait a thousand years or so to determine its survivability. To earthlings a thousand years seems to be a long time, but when the established long lifespan of an XTB is taken into consideration it does not appear so long. After the plants have established themselves the various animal lifeforms can be installed. In both instances the lifeforms will be designed to exist in the new environment.

The XTC performed this process many times over millions of years. If a life-form did not evolve as expected it was allowed to go extinct to be replaced by another. Regular natural catastrophes could erase all lifeforms. It would be logical to assume that undesirable lifeforms would not be replaced after a catastrophe. Accepting the possibility that the XTC replaced lifeforms after a catastrophe eliminates the need to stretch the theory of evolution to cross the boundary.

Much study has been done on unique life forms located in isolated geographic areas. This is based on the theory that a life form from an adjacent area migrated into the isolated area and commenced to evolve into a different species. In most cases the two areas are located thousands of miles apart separated by water. For the theory to be believable it must be assumed that both male and female of a species be transferred at the same time on wind or sea currents. The transfer of each member must occur within a few years, otherwise the existence of a possible mate would have ceased. Procreation with a resident species is not possible. The emigrant species may become nothing more than food for the resident species.

A more believable scenario is the tarraformers installed the unique life forms in isolated areas so as to observe evolution without contamination by adjacent areas. The use of isolated areas may lend credence to the location of the lost continent of Atlantis. The city is said to be circular in layout which suggest it was constructed on an extinct volcano. The theorized location is in the Atlantic ocean which is isolated from all land masses. This could have been the earthly headquarters for the tarraformers project. Even though the XTC is very smart they could not prevent the effect of plate tectonics which resulted in the destruction of their headquarters.

It can be postulated that the XTC tarraforming project is still in force today. This would explain the mysterious life forms being reported in great numbers as evidence of continuation of the tarraforming project. The science of denial is being used to protect existing theories. The fossil record shows that the earth has experienced many catastrophes in the past. Catastrophes occur at regular intervals. These do not include storms, minor volcanoes, and minor droughts. The minor events only serve to reduce the population in local areas which is predicted by evolution.

4-Extraterrestrial Life

Science chose to ignore and ridicule religious history. Now that the question of extraterrestrial life has come to the fore, it is left with the problem of explaining what happened before the scientific method. Although religious history was modified to enhance belief there is still nuggets of truth embedded. One passage from the bible supports the possibility of extraterrestrial life. This can be found in John 14:2-3 as repeated here.

In my father's house are many mansions: if it were not so, I would have told you. I go to prepare a place for you. And if I go and prepare a place for you, I will come again, and receive you unto myself; that where I am, there ye may be also.

This passage seems to imply another planet is being made ready to receive mankind. This could mean the XTC is aware that the earth is deemed for a major natural catastrophe in the future. This tends to support religious teachings that, if you believe in God and lead an exemplary life you will be transferred to the new planet. These conditions eliminate the need of having to save a major portion of the earths immense population. If the XTC is recognized as being extremely advanced, they still don't have the ability or desire to save the entire population of the earth.

This will be similar to what happened to Noah's clan. They had desirable traits that if saved would eliminate the need for the tarraformers from having to start over with their grand project. It must be realized that man is just another species in the tarraformers project. It would be nice to assume that man is a special species in the project.

Recently observational science has detected planets around other stars. There are three methods used in detection. The wobble detection method monitors the position of a star for variations in its location. If it moves in the viewfinder, there is a possibility that a planet is causing the motion. The presence of planets can be bolstered by the second method of detection which is the light signature of a star. If a planet is in an orbit that is parallel to the line of sight, there will be a decrease in the output of the light source at noticeable intervals. As the angle of the orbit to the line of sight increases light variations will become less detectable.

The third method proposed is the measurement of red shift of the stars light. This may be a useful tool if it can be shown that there is a regular change between a red shift and a blue shift. The planet must have enough mass to cause the star to exhibit a large velocity change towards the observer that can be detected as a red shift. If such is not the case it can be assumed the red shift theory has a problem. The red shift method becomes more problematic if the red shift is not periodic and associated with a blue shift.

The Search for ExTerrestrial Intelligence program is attempting to verify the existence of the XTC by listening for radio signals. Since none have been detected, it is popular to assume the XTC does not exist. A probable reason for non detection is that the XTC does not use radio for communication. A second possibility is they are using encryption schemes that hide the signals in the background radiation. More than likely the XTC does not use encryption, since man has only became sufficiently advanced to detect radio signals in the last one hundred years.

The detection of planets led to the question could these planets support basic life? Scientists are careful not to go so far as to purpose the possibility of advanced intelligent life. Support for the question of advanced intelligence was made by the occurrence of UFOs which could not be explained or ignored, but could be extrapolated as the existence of a superior civilization out there.

The monitoring of man's evolution has been going on for thousands of years, but man only became sufficiently advanced to recognize it since WWII. This is when the CIA began its debunking of UFO sightings.

Section-2
MAN'S BEGINNING

BILLIONS OF YEARS AGO an extraterrestrial civilization (XTC) started a tarraforming project on planet earth. This project resulted in installation of various life forms and observing their evolution. The members of this civilization will be referred to as XTBs. The XTB are referred to as pagan gods by religion and mythical gods by science. About 450 thousand years ago the XTC decided to send a mining operation to earth. You can imagine how this decision was received by the tarraforming XTBs who had been working on their project for millions of years. This clearly interrupted the path of evolution.

The conflict between the miners and tarraformers sets the upper limit to the age of ancient cities at 450 thousand years ago. This date was derived by Zacharia Sitchin in his numerous books on the subject. The theory of tarraforming proposes that a life form be installed and wait to see how it progresses. The XTC had no reason to build cities before the installation of the mining operation.

The exception may have been cities on the continent of Atlantis. This could have been the home base for the tarraforming project because of its isolation from the remainder of the planet. We will never know because all traces will have been erased by the process of plate tectonics. Ancient records speak of a city of circular construction. This tends to suggest the city was built around an extinct volcano.

Ancient mining cities were restricted to the area near the equator to avoid the ice sheet in the northern hemisphere. A comparison problem exist when the remains of the ancient cities of Sumeria are compared with those of Mesoamerican cities. The Sumerian cities are nothing more than rubble, while those of Mesoamerica and Indus valley still portray grandeur of construction. This may be explained by the Sumerian cities location in areas of dense population. The local population could have recycled the building material of the cities. In addition, a primary building material was sun dried bricks which deteriorates easily.

This leads to the erroneous assumption that the mesoamerican cities are much younger than those of Sumeria. In fact the dating of many mesoamerican cities has been skewed to support this assumption. This is commonly known as "shoe horning". If a scientist needs to assign the dating of an artifact to a desired time period, he only needs to claim contamination. One such instance is the study of the shroud of Turin.

2-First Mines

The first attempt at acquiring gold by the mining operation occurred in Mesopotamia. Many interpretations for the locations of the first cities were part of a landing grid. If the XTC could traverse space, they surly would not need landing beacons They had global positioning systems far superior to the ones of today. It could also be assumed that, if they did not have such systems they could not traverse space. Another point in question is evidence of chemical propulsion which is not viable for space travel. Maybe the illusion was the manifestation of another phenomena. The mining plan was to extract the gold from sea water. The attempt was not successful so the operation was moved south to Africa and other areas as an earthen mining project. The mining operation is the one event expanded upon in the cuneiform text that is used most to marginalize them.

Some of the arguments against the text are very compelling. The most asked question is, if there was mining where is the physical evidence? Mines can be classified as shaft or open pit. An advanced civilization would have ready access to equipment to allow both types of mining. An open pit mine is less labor intensive than a shaft mine. The evidence of open pit mining would erode so as to appear to be a natural landscape over a period of thousand of years. The natural process of erosion over thousands of years will eliminate any evidence of machinery. If erosion does not destroy the evidence, then alluvial deposits will cover it. It is possible that in the future archeologist may find evidence, but the question is will they recognize it as such.

Evidence of ancient shaft mines have come to light, but were attributed to man using primitive tools. It could be argued that normal geological processes erased evidenced of the entrance to a mine, but it is difficult to argue that geological processes erased entrances to all mines that must have been created over hundreds of thousands of years. Another possible answer for the lack of evidence is that the tarraformers were required to restore mining sites to a pristine condition the same as the EPA requires earthly miners. A strategically placed explosive charge can restore the entrance to a mine to a pristine condition easily.

If a deep shaft mine is located the evidence of the XTC will be available. Oxidation and erosion would not be as prevalent in the shaft and more importantly the tool marks and vast size of the shaft will be evident. A possible argument against this scenario is that the XTC was not equipped to perform shaft mining, and required manual labor which became problematic at a later date.

One possible example of an open pit mine is in the Nazca area of Peru which is not subject to erosion. The theory is that the leveled mountain tops were the creation of landing strips. Alternate explanations for the area are listed as follows:

1. Landing strips are not logical when one realizes that craft that have the capability of interstellar space travel would not need a landing strip, much less a landing strip

miles long. Even the Apollo space craft did not need a landing strip.

2. The assumption of landing strips would imply that the ships must use aerodynamics to gain altitude. Observations of UFOs show that they are not reliant on aerodynamics.

3. The other possibility is the supposed landing strip is the evidence of mining. This can be justified by the fact there is no debris found at the base of the leveled mountains, and it would be difficult to imagine that a primitive people could level mountain tops with primitive tools even for a religious purpose.

4. Another anomaly for which there is no reasonable explanation is the band of holes in the same area. These holes are small and in a narrow band that extends over a considerable distance. They appear to be the result of a sampling operation. The explanation given was they were used for grain storage.

5. The theory that the lines were irrigation channels is not reasonable, since they have no depth or controlled elevation, plus there is no water source.

6. Another explanation for the lines is that they were paths used for religious religious pilgrimages. If they were used as paths they would have been worn to a significant depth. No significant evidence of religious artifacts and human habitation exists.

7. The explanation of the lines as well as the pictographs on the plain is an effort of the native population hoping to encourage the XTBs to return.

3-The Miners

The two XTBs in charge of the operations on earth were half brothers. They will be referred to as XTB1 and XTB2 while their father who remained on the home planet will be referred to as XTB0

(GOD). XTB1 was placed in charge of overall operations on earth. XTB2 was in charge of the mining and other operations.

In recent times a rift is beginning to form between religion and science. The rift revolves around the definition of God. Although the evidence of ancient aliens is overwhelming religion has to maintain its support of belief in a single God which was initiated with the advent of monotheism. Since XTB0 was responsible for everything that transpired on earth, could he be thought of anything less than God? It is difficult to prove or disprove the unknown if questions are not allowed.

During the first 144 thousand years of the mining operation the ordinary XTBs performed grunt labor. It was at this time the XTBs working the mines commenced to complain. The result was the first labor strike. XTB1 brought the problem before the council of twelve. It was suggested that a primitive worker be created. XTB2 stated that the tarraforming project had access to a creature with the locomotion skills required by a miner, but not sufficient intelligence. The decision was made to increase the creature's intelligence level.

The creation of a primitive worker suggest that at this time shaft mines were the reason for the XTBs discontent. The work would be more demanding than operating more advanced equipment used for open pit mining.

It was not long until the new worker gained the attention of XTB1. XTB1 directed XTB2 to create some of the black headed workers for headquarters. Distinguishable characteristics, such as race, require only a minor tweaking of the DNA as proposed by modern scientist. The ninety percent of DNA not yet decoded is claimed to be junk. This junk contains the blueprint of a life form and not the ten percent that contributes to minor characteristics discovered by modern science. As time passed workers with different characteristics were created for different locations. This led to the many human races of today not evolution which could not be as specific as intentional modification. Having different races in disparate geographic locations made it easy to separate mutations from cross breeding as the result of migration. Scientist purpose that race came about via migration and evolution and dare not contemplate the possibility of intentional modification due to academic correctness.

The XTB spawned many offspring over thousands of years. Some of these were of Divine lineage while some were half Divine and half human known as demigods. These offspring although not directly associated with mining required a useful endeavor. They were designated kings and given the task of managing a human population which would produce food and build cities.

The existence of demigods may explain why ancient populations supported a line of inheritance in an effort to retain a linkage to the divine. The divine lineage became less obvious with passing generations. Although the demigods lived many human generations their offspring's lifespan decreased over time until it was little more than that of humans. Even today man is still possessed with the desire to leave any earthly possessions and knowledge to his offspring.

Much has been said about man's warlike tendencies. This should be expected since the ancient texts are replete with accounts of XTBs fighting among themselves before the advent of man. After the advent of man the XTB would use his kingship to wage war against that of his adversary. This is a possible explanation for the destruction of Sodom and Gomorrah. The bible states the reason for the destruction as punishment of sinners.

Ancient Indian text speaks of aerial battles that may have been between opposing XTCs. It can be assumed that earth's XTC won because the tarraforming project continued without noticeable interruption. War over territory and resources seems to be a common thread in the spectrum of life. The tendency can be seen in ant colonies as well as the strangler fig in the plant kingdom.

4-Howwhowhenwhywhat

Today the producers of video documentaries are asking the HWWWW questions. The productions use the existence of ancient structures that can not be explained as the creation of man with the aid of the XTB. The XTB are referred to as ancient astronauts that designed the structures and helped man with the construction. Many feats of construction could not have performed by ancient man. It is more than likely that the construction was achieved by the XTB with the use of advanced equipment. The only area conceivable for mans participation is trash removable. Remember these were the mine workers which were one level removed from modern man.

One instance that eliminates man as the prime contractor of a megalithic structure is the quarrying and placement of stones weighing hundreds of tons. He did not posse the tools for quarrying and definitely no means for placement. The benefit from the use of large stones would be negated by the effort required for their use. The effort expended by the XTB in manipulation of large stones must have been negligible. The dating of these ancient structures have been skewed to allow modern man to participate in the construction. Remember the miners had 450 thousand years to achieve the construction. Modern man was still using stone tools 10 thousand years ago. Building megalithic structures were of no significance to the XTB since they possessed the advanced equipment to achieve the task. They did not want to rough it, so they tried to achieve the comforts of their home planet.

The theory proposed was that the ancient constructions were left as a message to aid mankind. This would be backward thinking of the XTC since it would be more beneficial to man to leave the equipment used along with a user's manual. The functions attributed to many of the structures are defined by man's current knowledge which has many holes. The construction of megalith structures ceased with the advent of monotheism. This is when the majority of the XTBs were transferred to the home planet. The only advantage bequeathed to mankind was intelligence which allowed him to ask the question HWWWW.

5-End Of Mining

When the mining operation was terminated the offspring of Divine lineage were allowed to transfer to the home planet. These offspring had no connection to the home planet, so some chose to remain on earth which they considered home. When the tarraforming project was fully restored those of Divine lineage who chose to remain on earth were forced to abandon earth.

Only a very select few of the demigods mentioned in the bible were allowed to transfer to the home planet. The remaining demigods composed the group of biblical patriarchs and kings stated to have exceptionally long life spans. They tended to guide man's evolution for many years.

One can only guess at the reason for termination of the mining project. It could have been that the required amount of gold was obtained or the supply was inadequate. This tends argue against the supposition that the XTC will return to confiscate earth's resources. If the quantity of resources were sufficient or of the required type they would still be here. In addition, earth's population has depleted accessible resources to a major degree. We don't know when the mining operation was terminated but we do know it was before the end of the last ice age about twelve thousand years ago. This is when the tarraformers began correcting the mess made by the mining operation.

The first step was to transfer non essential mining XTBs back to the home planet. This could explain why many ancient cities were abandoned. The second step was to treat the mine workers as a newly installed life form. This required a deviation from the original concept of letting new life forms evolve without intervention. This required additional advancement of the intelligence of the workers beyond that required by the miners. This new creation was desirable since it would have the ability exist without assistance.

The modification of the primitive humanoid resulted in the creation of what the bible refers to as the Garden of Eden. The garden of Eden can be thought of being similar to today's game preserves or laboratories. More advanced forms of the mine workers were created. The new creations had increased intelligence and a free will. They were created as both male and female to allow the propagation of the species. Some time later the new beings were transferred from the laboratory to the local environment. The contrived reason given by religion for the transfer was the result of man's disobedience to the XTB. At this point in time we should commence referring to Adam an Eve as man and woman.

The mine workers were probably what we call Neanderthal. As time passed part of the theory of evolution is assumed to be proved to be true. Since the new version of man was more advanced than the mine workers, it was not long before the new version became the dominate species.

In addition to the Neanderthal serving as mine workers were also food producers. When the mining operation was terminated they were no longer needed. They did not have sufficient intelligence to exist without supervision. They had to resort to becoming hunters and gathers in order to exist.

Another possibility for the demise of the Neanderthal is the great flood myth from biblical sources. Although no definite time for the event has been established, and the occurrence of the event has not been validated either, the event remains as a part of human history. A possible time for the flood is about 13000 years ago when it is suspected that the earth experienced an undefined catastrophe. This is when it is generally accepted as the time of the extinction of the mega fauna. This could also be when the Neanderthal became extinct. Maybe the extinction of the Neanderthal is what religion refers to as God's elimination of sinners.

Over a period of a few thousand years the population of modern man increased rapidly. The second step in man's evolution was the transfer of knowledge that would allow the formation of self sustaining civilizations. XTBs managing a population of workers were given the powers similar to today's mayors and governors. These XTBs are what modern science call mythical gods. Religion refer to them as pagan gods. A priest hood was established to act as an intermediary between the XTBs and the population. The priests were tasked with transferring the knowledge of engineering and agriculture to the new beings as well as insuring they continued worshiping their XTB. In addition the priest was instructed to disseminate moral values which allowed the population to coexist within itself.

The knowledge transfer does not seem impressive when compared to the modern definition of knowledge. It did consist of a means of production of agricultural products that could be used as food as well as domestication of animals that could be used as motive power and food. The engineering emphasis was placed on the construction of massive structures with smaller stones which could be managed by primitive methods. In addition the construction of irrigation systems for food production was instituted.

A singular instance of knowledge transfer was the implementation of the calender. This occurred in 3760 BC for the Jewish and 3114 BC for the Mayans. The Mayan calender has received much attention in recent years because it ended on a significant astronomical date. Some has said that the date is not significant, it is just where the engraver ran out of space. It could also be assumed that the engraver started with an important date and carved in reverse order to a less important date. The calendar started in 3114 BC which must have been associated with an important date.

If Dec 21, 2012 is a significant date it must have been transmitted to the Mayans by the XTC, since they did not have the ability to compute astronomical events so far in the future. If they did have the ability they were using equipment supplied by the XTC.

Modern science proposes the reason for the development of the calender was to allow the farmers to know when to plant their crops. This assumption is not reasonable since the farmers could not read and had no access to the calender. Even today the calender gives a ballpark time of spring, but farmers wait until trees commence to bud. A tree's budding is a more reliable indication of the beginning of spring because it takes into consideration the local environment. For these reasons one must ask how and why was the calender created. Part of the knowledge transfer by the XTC was a method to mark the passage of time without which there could be no civilization.

After the XTB left earth the priest had to develop knowledge of events which was formed into beliefs. These beliefs were basically what we today refer to as theories. The theories were never proven, but individuals were forced to believe they were true by threats of force instead of empirical evidence. Religion never made any attempt to explain the beliefs in worldly terms since it was thought that its beliefs were supreme. It was mistakenly assumed no one would dare question a belief.

6-Scientific Method

About 500 years ago the intelligence and free will granted to man began to work its way to the surface. This time is quoted as the beginning of the scientific method. The scientific method proposes that conclusions not be based on belief but proof. This became a problem for the priest because up until this time they were the dispensers of controlled knowledge.

In the early days of the scientific method the church used various forms of force in an effort to repress it. The most misguided effort to force religious beliefs was the Spanish inquisition . This only served to reinforce the scientific method. Another instance of religion trying to suppress knowledge is that of the Spanish explorer's destruction of the Mesoamerican civilizations. The Islamic religion still use various forms of force to enforce its beliefs.

The discovery of Jupiter's moons by Galileo in the early years of the scientific method created immense problems for religion. It could no longer support the belief that the earth was the center of the solar system. The church in an effort to squelch his findings forced him into home arrest for the rest of his life.

Then Charles Darwin purposed the theory of evolution. At this time the church had lost its grip on scientific thought. His theory states that genetic mutations can be transferred across generations of a species. If a mutation is conducive to survival it will be transferred to the next generation. This theory allowed scientist to purpose that man evolved from apes, although the theory does not allow such. This incensed religion which resorted to demonizing the theory in the eyes of its members. Science did not utilize the Sumerian assertion that man was created from an existing humanoid which was not an ape.

The humanoid was more than likely a creation of the XTC and not an evolution from apes. Modern man was created from this base. Today one of the most contentious subjects between religion and science is whether evolution or creation is taught in schools.

If the tarraforming theory is accepted both theories can be shown to be true. Early man was created from an earlier life form by the XTC. This was not necessarily the ape because the theory of evolution does allow a species to evolve into a divergent species. Science has stretched the theory so as to allow new theories to develop. Any point of divergence can be shown to be a point of insertion by the XTC. Modern man was created from early man by the XTC and did not require insertion like other life forms. This would explain why science has not been able to find the missing link in the fossil record.

The theory of evolution does not allow for mutations across catastrophic events. It is probable that the XTC waits for the dust to settle after a catastrophe and then installs new life forms suited to the new environment. Species preceding the catastrophic event are seldom replaced, the alligator being an exception. An example of stretching the theory of evolution is that birds developed after a catastrophic event from feathered dinosaurs. A more realistic explanation is that the XTC liked feathers and decided to install the many bird species after the catastrophe.

The animosity between religion and science has been more detrimental to man's moral evolution than can be imagined. Religion is based on very ancient accounts of events in the very distant past when man was much less advanced. Much knowledge of events was transmitted verbally since writing was in its infancy. The problem for the scientific method is that it does not apply in these instances; therefore science ignores the event or goes so far as to ridicule it. This goes back to the conflicts at the beginning of the scientific method.

When science encounters ancient information it cannot explain the information it relegated to the realms of religion. One such example is that of Roman baths which is not so ancient. The purpose of the purification baths was not religious, but an attempt to eliminate the bodily stink and transfer of disease. Another event relegated to religion is the communal meal. The meal was an enticement offered to the hungry members of a group to insure that they attend a meeting for information exchange. We do this today.

The only written record that we have of the ancient past is Jewish religious texts, the cuneiform texts of the Sumerian civilization and the Sanskrit texts of ancient India. Even though the religious texts were created to fit a particular point of view they are based on bits of truth. Archeologists dig up ancient artifacts with the labor of students. This is a very demanding endeavor. Artifacts are then relinquished to the armchair archeologists who interpret it according to modern knowledge "guesses". Theories are then formulated to explain the results.

Be aware that a theory is a belief waiting for proof. In many cases the proof is nothing more than a guess that is accepted as truth based on the prominence of the author. Archeologist expend much effort trying to suppress the discovery of artifacts by non-archeologist. This is effective only in those instances where the possible existence of artifacts is contained in particular state controlled sites. It does not prove to be effective when applied to bulldozers and chance. Reference the Dead Sea scrolls.

If religion and science had worked together the knowledge of the XTC tarraforming project would have already been recognized. Acceptance of the theory will have relieved religion of trying to prove the unprovable and science of explaining the unexplainable. This in turn may have hindered the evolution of man because scientist would become nothing more than a priest. There are responsible scientists secretly working to reconcile religious text scientifically. They have to work secretly to avoid retribution by their peers.

Man's knowledge has advanced exponentially in the last 500 years. This advancement can be broken into two parts. The first is measured by the scientific inventions that have increased the lifespan and made existence easier as evidenced by the population explosion. Second science is no longer impeded by ancient religious belief.

Recently observational science has detected planets around other stars. This led to the question could these planets support basic life? Scientists are careful not to go so far as to purpose the possibility of intelligent life in this respect they are similar to priest. Support for the question of intelligence was made by the occurrence of UFOs which could not be explained or ignored, but could be extrapolated as the existence of a superior civilization out there.

7-You Decide

The investigation of any event requires the answer to the standard questions "how,who,why,what when". Acceptance of the tarraforming theory will answer the question of who built the ancient constructions and who pilots the UFOs. All the other components of the queries will require new theories which will replace current theories. The new theories will remain guesses until the XTC decides to make contact or we stumble upon them.

Since it is difficult for man to envision an intelligence that may be millions of years more advanced. A new discipline will need to be formed to allow the detection of the fingerprints left by the gods. The most used debunker phrase is "if they exist why haven't they made contact". If it is assumed the XTC can set in orbit and monitor TV and radio broadcast to keep abreast of happenings on earth, why place boots on the ground? A recent rumor trends to support this assumption. The XTC discovered that the XTBs tasked with monitoring TV broadcast were spending too much time playing fantasy football.

Section-3
RELIGION

THE LAST ICE AGE ended about twelve thousand years ago. Any event occurring before this time is known as prehistory. Almost all we know about prehistory came to us from the cuneiform text bequeathed to mankind by an extraterrestrial civilization (XTC). The XTC had been conducting a tarraforming project on earth for millions of years. The individual members of the XTC will be referred to as extraterrestrial beings (XTB). This essay will propose that the supreme entity of the XTC is what we call God. Lesser entities called XTBs will be shown to be what religion calls pagan gods and science refers to as mythical gods. Some of the lesser XTBs are referred to as angels depending on their assigned task.

Chapter one of genesis in the holy bible explains the time line for the creation of the solar system and not the universe as some think. The time line is laid out in days. This is the most used point of argument for the debunkers. It must be realized that when the knowledge was transmitted to man one of the measurements of time understood by most was a day. Early man did not have any concept of billions and millions. From this it may be assumed that a day was used to explain time segments of events which could have been billions of years.

The second verse of genesis implies that the planets were present before the ignition of the sun as stated in verse three. Verses four through ten deal with the geological processes that affected the early earth. Verse eleven can be seen as the beginning of the tarraforming project's installation of life forms. This assertion presents the possibility that there are many solar systems in the process of forming but cannot be detected because their sun has not ignited or inadequacy of detection equipment.

Some 450 thousand years ago the XTC inserted a mining project into the tarraforming project. It was decided to create a mine worker from a humanoid life-form in the tarraformer's inventory. The humanoid had the required locomotion skills, but not the intelligence. He was not an ape. His intelligence was advanced so as to be able to perform as a worker. This time span is documented in the cuneiform text of ancient Sumeria.

After the mining project was terminated the tarraforming project had to clean up the damage done to their project by the mining project. Rather than destroy all the intelligent workers that were created, the decision was made to treat them as a newly installed life form. The XTC realized that they advanced the process of evolution by millions of years, but the life-form would not be able to exist with its current intelligence level. That resulted in its intelligence being advanced even further.

This resulted in the creation of what the bible refers to as the garden of Eden. The garden of Eden can be thought of being similar to today's game preserves or laboratories. More advanced forms of the mine workers were created by genetic engineering. The new creations had the ability of free thought, free will and procreation. Some time later the new beings were transferred from the laboratory to the local environment. This was more than likely a method of promoting the new species. The biblical reason given for the transfer was the result of man's disobedience or sin.

After Adam and Eve were kicked out of the Garden of Eden they had two sons. Confrontations erupted between the two which resulted in Cain's killing of Abel. God was still involved with his creations. When he saw through Cain's alibi for the killing, he banished Cain to the land of Nod. One must ask, if Adam and Eve were the first humans who populated the land of nod.

As time passed the survival of the fittest posit of the theory of evolution proved to be true. Since the new version of man was more advanced than the mine workers it was not long before the new version became the dominate species. The mine workers were probably what we call Neanderthal who inhabited the land of nod

A Jewish priest using the life spans of the patriarchs computed the garden of Eden event occurred in 4004 BC. He made the mistake of claiming this is when the earth formed. However, the date seems reasonable for the creation of modern man. From this time and until the Jesus event the XTC was actively involved in man's evolution.

2-Flood Myth

Shortly after the creation of man a significant catastrophe occurred. This event is disavowed by science, but it is still there and they cannot prove it or disprove it, so label it as myth. The worldwide flood mentioned in many religious texts could have been the result of the Planet Of Crossing(POC) passing close enough to earth so as to create massive tides. The biblical account of the flood speaks of an extended period of heavy rain as the cause. This claim and others will be analyzed in following list.

1. Heavy rain was observed because of association with high tides caused by the gravitational effect of the POC. The high tides caused the flooding, not the rain. If rain had caused the flood, one must ask the question where did the immense amounts of water come from and where did it go.

2. The size of the boat was much more than was feasible in that point of time. The XTC may have collaborated in the construction. Even the immense specified size was not enough to accommodate the prescribed cargo.

3. The time of construction specified in days may have been another instance of the ancients of assigning longer time periods to the day. Even today earthly scientist can predict events years in the future. Surely the XTC would have equal abilities.

4. The directions to save a pair of each animal is unrealistic. This does not take into consideration the thousand of specifies of animals. These species could not coexist in close proximity to each other even in cages.

5. To complicate the collection process even further, one must consider the logistics required to capture wild animals. Even if they are captured how is a food supply for each is amassed. Maybe it was assumed that the time span of the catastrophe was short enough as not to require a food supply.

6. Since the directions do not specify the animals to be saved, it can be assumed that only domesticated animals were included. This would be a logical direction of the XTC, since it would preserve man and his dependent animals. It would prevent the XTC from starting over from scratch with their project.

7. This would explain the extinction of the mega fauna since they were not on the boat. This tends to marginalize current theory that primitive man was responsible for the extinction using primitive tools. This could also explain the extinction of Neanderthal.

8. 8. Since the flood myth is included in many unassociated civilizations beliefs it must be accepted as fact by science. Science should spend more time trying to validate the event rather trying to ridicule it. This is an instance of a natural catastrophe that did not leave any recognizable physical evidence.

3-Early Man

The bible speaks of minor XTBs having intercourse with the new earthly women. The result of such a union was a being known as a demigod. That is half human and half divine. These demigods were the first priest and kings. After the XTB no longer had contact with humanity the priest and kings strove to maintain the divine linkage. Over the following generations the divine contribution to the offspring becomes less and less. This did not deter the priest and royal families from claiming a divine lineage.

The project management XTBs formed priesthoods from the demigods whose members would act as intermediaries between a local XTB and man. The priest would also transfer knowledge to man. The priest insured man would worship his local XTB (god). If man refused to adhere to the instructions of the priest, bad things could happen to him as will be seen later. Some instances suggest that the XTB performed acts for his own amusement. As time progressed the priest were given more power to influence the evolution of man. This decision was made by the XTC to allow man to evolve without further intervention because they wished to withdraw to the point of being observers.

It wasn't long before it was realized that the priest could not exert absolute control over the population. This is when it was decided to initiate king-ships from the demigods, who had the power to wield physical force on the population. This was in direct opposition to the priest who could only wield a moral force. It is logical to assume that where there is opposition there will be conflicts between the priest and kings.

4-Ancient Priest

One of the priests of ancient times was Abraham. In his later years he still did not have an heir. His wife Sara agreed to allow him to foster a child by her hand maid. The child was male and named Ishmael. After the birth of Ishmael the XTB heard Sara lamenting the fact that she could not give Abraham an heir. The XTB told her that she would bear an heir. Sure enough she did give birth to an heir who was named Isaac. In a short time a confrontation arose between Sara and her handmaid as to the order of the inheritance of the sons. Sara convinced Abraham to banish the handmaid and Ishmael. Ishmael was to become a Muslim and Isaac became a Jew. So now you know why Muslims and Jews can never coexist.

Abraham was told by an XTB that he should sacrifice Isaac. At the point of executing the sacrifice Abraham was told to cease. The reason given in the bible was the order was God's way of testing

Abraham's faith. An alternate explanation is the original order was given by a rival XTB and was stopped by Abraham's XTB .

Another instance of an earthly entity being affected by the will of the gods is that of Job. Job was a wealthy tribal leader. One day a rival XTB over a few beers engaged in a contest with Job's XTB . Job's XTB stated that Job was a dedicated servant. The rival XTB stated that he could change that. Job's XTB said show me. The rival XTB set about to do so. Very bad things began to happen to Job. After much suffering by Job the contest was terminated and Job regained a semblance of his previous status.

The word sin is used to emphasize acts performed by the populace disapproved by their XTB . This could be applied to the entire population of a rival XTB . This resulted in many of the ancient wars. After direct contact with the XTC ceased man continued the war like tendencies learned from the rival XTB s.

The best known claim of a sinner's punishment is that of the destruction of Sodom and Gomorrah. Abraham's nephew Lott was a resident of Sodom. It was decided with Abraham's insistence that Lot was worth saving. So angles (XTB) told him he needed to pack up and get out. Lott was told to evacuate and not look back. His wife did look back and is said to be turned into a pillar of salt. This statement was accepted by the uneducated who could see salt formations that took on the appearance of a human form.

The pillar of salt has been shown to be a mistranslation of the original Hebrew of a pillar of vapor. So why change the interpretation after thousands of years. Maybe Lot's wife was dragging up the rear and was still in the blast zone. The question must be asked where did Lot procure enough wine to get drunk? The answer that he carried it with him is not realistic since the weight and bulkiness of a container would not allow enough wine to induce drunkenness.

Since the angels(XTB) knew the event was going to occur, it must have been self directed or the atmospheric explosion of a meteorite. The atmospheric explosion assumption is mandated since no craters have been found. A self directed destruction of the cities may have been executed to erase an undesirable path of evolution.

The Jewish civilization can be thought to have commenced with the creation of Adam. In these early times God made the covenant with Abraham that his descendants would always have claims to the land of Israel. The XTB in charge of the Jewish population appears to be the most efficient of all the XTBs in charge of populations, since the Jewish religion remains barely unchanged over thousands of years. The following is an appropriate time line for the occupation of the land known as Israel.

1. Sometime after Abraham his descendants migrated to Egypt because of a drought. At this time they no longer had the direct assistance of an XTB . The Arabs reoccupied the vacated land.
2. When Moses led the Jews out of Egypt they reoccupied the land by many bloody wars.
3. After the Jews were exiled to Babylon by a more powerful military power the Arabs reoccupied Israel.
4. When Babylon released the Jews they reoccupied Israel by migration and war.
5. In the 1200s the Arabs and Jews shared occupation of Israel. This is when the crusades embarked on an effort to remove the Arabs from control of the Holy land. The crusades did nothing but increase the animosity between Arabs and Jews.
6. At the end of WWII the state of Israel was recreated. They reoccupied the land as the spoils of war, since the Arabs were seen as supporting the Axis.
7. The Jews have been able to maintain occupation of Israel through various wars since WWII. Some people assume the reason for Israel's continued existence is the result of the original covenant. It can be more realistically attributed their tenacity as a population retain their beliefs in their past and support of Christians.
8. The support of Christians can be seen the same as children supporting their parents.

5-Monotheism

A few thousand years after Adam was created the decision that man could exist without intervention led to the transfer of XTBs to the home planet. To explain the lack of local(pagan) gods the existence of an unseen God that resided in heaven was instituted.

The XTBs associated with the mining project had been transferred thousand of years earlier. The last XTBs to be transferred were members of the tarraforming project that were responsible for local populations of man. The existence of the populations were now under direct control of the priest.

The Jewish religion evolved during this period. Most of the book of Genesis was also composed during this period. Portions of the book relied on the Sumerian texts which were referred to as olden text.

Priest in other parts of the world developed religions at this time. These religions have little in common with the Jewish religion. This implies that they were formed from knowledge from their departing XTB and some creative thinking by the priest.

Few religions make any solid reference to an extraterrestrial influence. The most extensive texts pertaining to the XTC is the Sumerian and those of the Indus Valley. However, a passage in the bible that alludes to the possibility of the the existence of the XTC can be found in John 14:2-3 as repeated here.

In my father's house are many mansions: if it were not so, I would have told you. I go to prepare a place for you. And if I go and prepare a place for you, I will come again, and receive you unto myself; that where I am, there ye may be also.

This passage can be interpreted in many ways. One such way is that the XTC is tarraforming other planets. In the event of a catastrophe those members of the population with the desired genetic traits will be transferred to another planet. Similar to what transpired with Noah during the flood.

This passage was more than likely the result of a conversation with an XTB. The priest used it to instil the belief of an after life and reincarnation in an effort to entice their followers to continue to follow. Neither of these tenets make any sense when the vast numbers of man that have lived and died over the thousands of years with varying degrees of intelligence is taken-ed into consideration. One must ask of what use these souls have for the XTC when many souls are created every day by the process of procreation? The tenets do have the positive effect of creating a belief system that allows a population to exist within itself.

6-Jesus Event

A few thousand years after the institution of monotheism the hi priest found it difficult to get their followers to worship an external unseen god. The solution was the creation of a demigod deemed to be the son of God whose task was said to be a savior of mankind. This was an attempt to bolster the belief in an external unseen God. Almost nothing is known about Jesus's formative years. An American Indian belief that a star child is transferred to heaven at age six to be educated by his star father could be the answer.

In his adult years his teachings became more and more popular. Some priest and the Roman emperor saw Jesus as a threat so they crucified him. The XTB did not have anything to do with the crucifixion, but they did not try to stop it either. The crucifixion had a desirable effect. The crucifixion of Jesus allowed his followers to create the Christian religion which is still in existence today.

Judaism is a religion that dates back to ancient times. Christianity came into existence with the crucifixion Jesus and retains many of the teachings of Judaism. For the first three hundred years Christians were persecuted by the Roman Empire. Then the roman emperor Constantine claimed to have seen the sign of a cross before a battle. His vision may have come due to a late night visit by an XTB (angel) or he may have realized that it would be easier to remain in power with a large following. Regardless of the reason the result was his naming of Christianity the Roman Empire's state religion. Henceforth, all branches of the Christian religion will be grouped under the term "the church".

About 600AD the Islamic religion came into existence. The Islamist claims the same Patriarchs as Judaism and Christianity, but this is where the similarity ends. This is the first time we hear of the conflict between Isaac and Ishmael, the sons of Abraham. The Islamist realizes that if you can control the population, you can control the world. See Iran as an example. To control the population one must eliminate all none Islamist, especially atheist. The Islamic religion has factions the same as the Christian religion. The Sunni is in opposition with the Shiite.

7-Predictions

In the beginning the priest promoted the wishes of the XTB. After the XTC abandoned earth and the priest could not perform the miracles of the XTB they devised other methods to insure control over the population. This became known as predictions which amounts to control through fear.

The occurrence of natural catastrophes has been used by religion to impress on their followers that their sins caused the catastrophe. A reason for the punishment claim was the population ceased to worship God, so in a vindictive effort he is said to have delivered the punishment. God had nothing to do with the event. When natural catastrophes are used for a prophecy no time can be given. When a catastrophe occurs the prophesier or his followers, even after many years, can say we told you so.

When natural catastrophes did not occur at regular intervals the profits turned to prophecies based on social events. Although the population repented for their sins it had no effect when a more powerful civilization decided to take control. This happened many times to the Jews, who have survived to maintain their beliefs. You would think that by this time that the Jews would have realized that their XTB was no longer able to uphold any covenants. The Jews belief system is so strong that it has weathered many disappointments. The belief system is primary to survival not a convenient of an XTB that is no longer present to enforce it.

The publisher of the plain truth magazine in the 1950s predicted that WWIII will occur in the Middle East. The WWIII prediction probably will not be realized since the only contribution to civilization the Middle East has to offer is oil. The common saying "you can't eat oil" is significant because the area will never be able to produce enough food to support a large population. Since most governments in the Middle East are formed around tribal chieftains who lord over camel drivers, they can never cooperate long enough to become a threat. Rather than WWIII a local war could erupt. When Iran secures a nuclear weapon it will use it on Israel. Retaliation will result in Iran being turned into a wasteland for thousand of years. The only thread that holds the Middle East together is oil and the Islamic religion. Oil reserves are being depleted at a rapid rate, so in the future the only thing left is religion. The Islamic religion is being threatened by modern communications with the outside world.

Religions have worked very well in advancing the evolution of mankind. Since the common man in the past could not read or write the priest were required to transmit knowledge. The translation of the bible from Latin to modern languages decreased the power of the priest. However, biblical texts are still in an archaic form which requires a priest to interpret them for the masses. Religion is based on belief and those who do not believed are termed atheist and cannot be saved.

One must ask, if the XTB were so advanced why would they spend so much time on religion. The ancient texts speak of the religious observances of the XTB, even before the advent of modern man.

Maybe this was a method used by the XTC of controlling the XTBs in the field. It worked so well the method was used to control the populations of man. It seemed to work for a few thousand years, but then the intelligence granted man began to take hold. When extremely religious countries are compared to those that are less religious, it can be seen that the less religious ones have done more in the enhancement of human existence. The single event that created more atheist than any other was the advent of the scientific method which requires proof and not belief.

8-Scientific Method

Following the crusades some members of the population began to doubt the claims of miracles made by the church, since none such have occurred after Jesus. The reasoning was that the miracles were fabricated or elaboration on common occurrences. Claims that catastrophes were God's punishment for man's sins began to lose credibility with increased knowledge of the natural world.

In the 1500s persons with scientific ways of thinking began to question the church's assertion that the sun revolved around the earth. Such persons could be declared a heretic which could result in torture or death if they would not recant their beliefs. The death sentence involved being burned at the stake. This period of time became known as the Spanish inquisition.

Galileo whose discovery showed that the sun does not revolve around the earth recanted but was subjected to house arrest for the rest of his life. This punishment was probably due to the fact that the church knew he was correct but it did not know what to do about the discovery. It had to take him out of circulation among the scientifically inclined in the population.

It was during this time the church destroyed the mesoamerican civilizations that it felt was in opposition to its teachings. In addition, the church realized significant monetary gains. It melted down the golden priceless artifacts which opposed its belief system.

This point is recognized as the beginning of the scientific method. The scientific method requires claims to be proven. It also allows claims to be modified or disproved. The problem arrives when the claim occurred so far in the past that there is no way to prove it true. In such cases the scientific method requires the claim to be declared a myth and the church requires one only to believe that it is true. It must be realized that religious knowledge contained in the bible contains bits of truth. Not all ancient texts were included in the bible. Various reasons dictated why some texts were included while others were excluded.

The XTC knows that an astronomical event will occur that will destroy all life on earth, it has happened before. This information was transferred to the ancient priest in a fuzzy form. Since the event was far in the future and the priest could not communicate events far in the future no specific date was noted. The book of revelation of the bible retains this knowledge in a fuzzy form that is intended to be scary to sinners.

The holy bible states that in the end time God will return and save all believers including the live and dead. This does not make sense, if it is assumed that dead of past centuries have already transferred to another dimension. It is more than likely that a few thousand beings will be saved if they posses genetic traits the XTC deem worth retaining for their next project. The minimum number to be saved is two, male and female. A possibility is that those saved may be used to populate another planet. This assumes that the XTC is pleased with the evolution of man.

It will be beneficial for religions to commence the study of UFOs. They have nothing to loose since the scientific community is opposed to religion. This would allow the creation of theories to support religious claims of ancient times. The study of UFOs would place the scientific community in the position of having to defend it's assertion that the phenomenon is imaginary and that the contents of the bible is all myths. It would also allow more scientists to study the problem without fear of grant refusal and peer ridicule.

Section-4
SCIENCE

THE SCIENTIFIC METHOD IS said to have started with Galileo's discovery that the sun did not rotate around the earth. He was treated very severely by the church which is resented by the scientific community till this day. Since the church's in-question could not get him to recant his theory he was placed under house arrest for the rest of his life. Probably the reason the church did no burn him at the stake was it knew his assertions were valid. In addition he was not your run of the mill heretic and his death would have caused repercussion. This could be put in modern terms as kicking the can down the road. His discovery were widely distributed upon his death. Most of his closest associates knew of his discovery but they kept quite due to fear of retribution.

Some time after Galileo a newly recognized scientist developed a leading principal of the scientific method. This principal is known as Ocam's razor. It states the most probable explanation for an event is the simplest one. This is a two sided principal. If no one tries to prove it true or false, it will be accepted as true. An explanation may be proposed to hide the lack of knowledge of the event or the possible fear of the consequences if it is true. This is the most effective implement in the Debunkers toolkit. Ocam's razor is mostly applied to events reported by non-scientist. Scientists generally report an event as a theory to be supported by other scientists who have a vested interest for support.

The scientific method states that a theory must be rigorously tested until it is proven true. If a theory cannot be proven to be untrue it may be accepted as being true. A problem arises when there is no method for testing the theory. The easy way out is to declare a myth for information that is very ancient, if it is modern declare speculation. If neither of these explanations are accepted just call it a hoax.

2-Theory Of Realitivity

The scientific theory that illustrates the best of the scientific method is Einstein's theory of relativity which resulted in the famous equation E=MC2. The theory has been proven consistent by many methods over the more than the hundred years since its inception. The equation implies that an object with mass cannot achieve a velocity greater than the velocity of light. Since its inception many scientist have devoted their life to prove or disprove the theory consistent.

With the development of improved observation technology scientist at the CERN observatory have determined that neutrinos can travel faster than the speed of light. This may not invalidate the theory of relativity since it is assumed that the neutrino has negligible if any mass. This leads to the question that if there is no mass to be converted where does the energy originate. Should the equation be restated as E=C2? In addition if the neutrino has no mass what is it. Maybe it is a cousin to light's photon. Although the photon is assumed to have no mass it is still considered to be a particle. This allows light to be assumed to be composed of particles and electromagnet waves. You can't have it both ways.

This supposition of two methods of behavior will have repercussion in the future. These two methods of behavior are required to support current models of reality.

3-Big Bang

In the early 1900s a prominent English scientist proposed the theory that the universe is in a steady state. The theory predicted no beginning and no end. In the mid 1900s a new theory found favor. This theory predicts a beginning and an end.

This new theory was developed as the result of an astronomer's observation that when the light from distant galaxies was passed through a prism, the elemental markers were shifted towards the red end of the spectrum. In an effort to explain the observation an analogy with the Doppler effect created by a mechanical sound source moving through the atmosphere. The Doppler effect analogy could then be used to determine if a light source is moving away or towards the observer. This discovery seemed to eliminate the need for the necessity of the cosmological constant in Einstein's equations. Scientist had been working years to derive a reason for the cosmological constant. Since the red shift theory seemed to supply the desired value the theory was enthusiastically embraced.

Since light from most distant sources was found to be shifted towards the red end of the spectrum, it was assumed that the sources were all moving away from the observer. This assumed the motion could only be explained as the result of an explosion. The explosion analogy led to the BIG BANG theory. This theory proposes that many billions of years ago all the mater in the universe was contained in what is called a singularity. It is hard to imagine that all the matter in the universe was concentrated into a single "glob" no matter how densely packed. For some unknown reason the singularity exploded. The result was the ejection of basic particles in all directions as described by a spherical shell. As time passed the basic particles began to coalesce into stars and galaxies.

As soon as the theory was proposed observational scientist commenced to try to prove the theory. Using the red shift it was soon realized that everything was not moving away from a known center. The problem was solved by assuming the explosion was not uniform which is a theory within a theory. The discrepancy could also be explained by matter between our position and the center of the explosion.

Observational calculations began to show that there is more matter in the universe than is observed. The explanation for the discrepancy was the concept of dark matter another theory within a theory. Dark matter could add credibility to the theory, if it is included as a product of the explosion. This would explain where the dark matter originated, but not what it is composed of. It would go a long way in explaining why there are black holes in the center of spiral galaxies. It would also explain how stars form, since it would be easier to postulate that the basic particles were attracted to the dark matter and not themselves. An effort to make star formation more acceptable rather than being formed from basic particles the dust component was added. No mention is made of how the dust originated or what it was composed of.

Another problem to arise was the fact that many objects were still moving apart at an increasing speed.

The concept of dark energy which was hypothesized to have come into existence sometime after the initial explosion was proposed to explain the discrepancy which inserts a theory into a theory.

Dark energy may be the power source utilized by UFOs. The previous statement was not intended as a joke.

Proof of the Expansion

If the universe is continually expanding there should be an exceptionally large void around the location of the initial explosion. If such a void does not exist, it must be assumed that matter is being generated at the initial location of the explosion or the explosion was not uniform in accordance with the discrepancy mentioned previously.

Galaxies viewed across the void should have red shifts significantly different from those viewed on the same side of the void as our galaxy. This seems reasonable since our galaxy and those across the void are moving in opposite directions, thereby creating an unusual red shift that could be interpreted as increased speed. This could be the explanation of the red shifts that suggest the universe is expanding at an increasing rate. If the center of such a void is not known, it is difficult to imagine how the age of the universe can be accurately computed.

Since the big bang theory is formed around the red shift theory which in turn is formed around an earthly sound wave analogy, it would be beneficial to both theories if it could be proved that the red shift is due to motion and not the bending of the light by the dark matter or absolute distance. Such a proof can be had by utilizing the hundred of years of photographic plates of distant galaxies.

This proof does not utilize the red shift to indicate motion. The proof relies on the knowledge that matter ejected by an explosion moves on different trajectories relative to the point of the explosion. Galaxies on the outer perimeter of an explosion will not be affected by the possibility of a non-uniform explosion. That is to say such galaxies would always appear to move apart. By measuring the distance between two distance galaxies over a period of years it can be determined if they are moving apart relative to each other. If no positive relative motion is detected then the validity of both theories will be in question.

Assuming the hypothesized dark matter was a component of the explosion it would be logical to expect that the gases would be attracted by gravity until enough mass was accumulated to form a star or planet. The stars would migrate into an orbit within the gravity field of larger clumps of dark matter thus forming galaxies. It has been determined that the center of most spiral galaxies contain black holes or dark matter.

Alternate Creation Theory

True to the scientific creed there is a group of scientist purposing an alternate theory for creation. This theory is based on multiple universes contained in membranes adjacent to each other. The theory has not been widely distributed to the popular press yet. Regardless of the discrepancies in the big bang theory it is more believable than the membrane theory. Then again neither of the theories may be correct and the steady state theory is all that is left. Both theories rely on the red shift theory for validity which may not be in itself correct.

4-THEORY OF EVOLUTION

Another famous theory that has been proven is Darwin's theory of evolution. His theory states that genetic mutations can be transferred across generations of a species. If a mutation is conducive to survival it will be transferred to the next generation. The theory does not allow a mutation of a particular species to to be transferred to a different species. That is to say procreation between species is not allowed.

The first problem caused by scientists was to purpose that man evolved from apes. This incensed religion which resorted to demonizing the theory in the eyes of its members. Science did not utilize the Sumerian assertion that man was created from an existing humanoid which was not an ape. It has ever sense been searching for a fossil that could supply the missing evaluational link between ape and man.

The second problem for the theory is its use by scientists trying to explain the appearance of the many species of plants and animals following a catastrophe. The problem is defined by the fact that there is no parent species to evolve from. Since the time spans following a catastrophe could be thousand of years, no life forms could subsist. This problem is solved by assuming that the tarraformers installed new lifeforms following the catastrophe.

5-What Is A Scientist

Today's scientists fall into two groups. The first is those who invent processes that enhance the ease of everyday existence. The second is those that propose the reason for our existence. Their problem is that some of the reasons given are not accepted. One such example is that man evolved from apes. The time line for modern man's evolution is too short to adhere to Darwin's theory of evolution and ancient texts states that man was created by the XTC.

In the 1800s universities began issuing a PhD degree. This entitled the holder to be known as a scientist. The following is a list of areas of employment available to a scientist.

1. Research and skilled projects in private industry.
2. Research and skilled projects in government.
3. Teaching at universities referred to as academics.
4. Lobbyist

Researchers and project scientist are given specific tasks. Their career advancement is determined by how well they perform the task. Many companies will reward a scientist with 100 shares of stock for a new invention that could be worth millions of dollars. The invention is retained by the company. These are the scientists that advance our knowledge of the physical universe. They invent processes and machines to enhance productivity which requires fewer jobs. The government scientists provide useful services in many areas that have not attracted the interest of private industry. One area where this is most evident is crime scene analysis. Data collection and evaluation is another area of employment. One must realize work published by US government scientist is generally tainted by political influence. One good example is the swamping of the airwaves with lead paint commercials sponsored by the EPA. In fact anything sponsored by the EPA is questionable. We need to de-fund the EPA and rename it APE.

The scientific community can be judged by its handling of the BP oil spill as a failure because of the governments influence. Now we have the Japanese nuclear problem. Although the scientist have no direct knowledge they are busily trying to get on a news programs with hopes of getting their scientist card punched. The intelligence quotient of Japanese scientist is far superior to that of US scientist.

Academics start their careers as teachers. To sufficiently advance their career they must apply for grants from the government. The easiest grant to acquire is for the study of a project the government deems necessary. Another type of grant is issued by the university for the Study of popular issues. If the scientist successfully completes the task he will be assured of getting another grant. The more grants the easier it will be to secure another grant. This is another way of looking at the common saying among scientist that one must "publish or perish".

6-Debunkers

Scientists tend to try not being favorably associated with religion because the old rift between science and religion. The mention of unexplained phenomena must be avoided because the senior scientist has learned that it is detrimental to their careers, if they show interest in any of the taboo subjects. The government reduces the scientific creed to mediocrity. One unexplained phenomena that the government insists that does not exist is that of the UFO. This is due to national security implications perceived during the cold war.

A scientist may be recruited to debunk information. Occam's razor is used to explain away the information without expending any effort. If one chooses to become a debunker of unexplained phenomena he will be viewed favorably by the senior scientist beholding to the CIA. Debunkers are generally scientist willing to debase the scientist creed to get their card punched and monetary gain. Debunkers are required by producers of video documentaries about unexplained phenomena who wish to appear fair and balanced.

Another path to name recognition and monetary gain is to secure contracts with special interest groups. This takes the form of research or serving as an expert witness in lawsuits and criminal proceedings. In such cases one is paid for results and not truth.

7-Bad Grants

One area where government grants can do more harm than good is that of the study of invasive species. The grants can have long terms and are given to scientists who have no knowledge of the subject. After the grant expires the invasion has progressed for years and no solution to the problem results.

Better results could be had, if the government would use the money expended on the grant to pay for a bounty on the species. The bounty system is very effective as evidenced by the elimination of the wolf in the 1800s. The effectiveness of a bounty system is enhanced by the fact that collectors of the bounties live in the affected environment and understand the species.

A non related bounty system has been initiated by the government to allow exposing company fraud by an employee. This eliminates the need for regulating authorities having to detect the fraud. The fallacy of such bounties is that it can be made by disgruntled employees pursuing monetary gains. This in turn costs the company and country many dollars. The media and politicians benefit from such bounties regardless of who wins. Stockholders of the company can be guaranteed to be losers. Of course, the government benefits from derived fines.

8-Searching For The Xtc

This research is done by those scientist not associated with academia or government. The exception is those involved in the search for extra solar planets. They are careful not to propose the possibility of intelligent life on these planets. In recent years science has developed equipment that can detect planets around other stars. Using the equipment astronomers have detected hundreds of planets. Only a few are of the composition and at the right distance from the star to allow life as we know it. There must be many with the required characteristics that go undetected with today's equipment.

Recently much effort has been expended to identify stars with planetary systems. It was observed that stars suspected of having planets seemed to wobble which suggested the tug of planets. The magnitude of a planet's tug on its star is determined by the inclination of its orbital plane with the line of sight.

If the plane is perpendicular to the line of sight, the star will appear to move left and right. If the plane is at a right angle to the line of sight, the star will appear to move in a circle. This is known as the wobble method.

Another method used is the transit method. This involves detection of periodic variation of the light from a star. This can be assumed to be the result of a planet passing between the observer and the star. If the plane is perpendicular to the line of sight, the star's light variation will be maximum. If the plane is at a right angle to the line of sight, the star's light variation will be undetectable.

The latest observations have shown that some stars exhibit a periodic red shift. This was interpreted as the tug of planets. This interpretation is questionable since planets do not emit visible radiation. Even if planets are present they do not have enough relative mass to cause sufficient velocity change of the parent star to produce a measurable red shift. It is more than likely that the passage of the planet dims the received light in such a way as to appear as a red shift. Such observations must also include an associated blue shift.

The Search for Extra Terrestrial Intelligence (SETI) project has been searching for alien broadcast in the radio spectrum for many years. Since no signals have been detected the probable reasons are the signals are outside the detection range or the XTC does not use the radio spectrum for communication. If UFOs were using radio for ship to ship communication, SETI would have inadvertently detected such transmissions. Since no ship to ship communications by UFOs have been detected the debunker will interpret the fact to show they do not exist.

9-MODERN THEORIES

To emphasize how theories can be developed while on the inside looking out some popular theories must be examined.

The theory of evolution has been stretched beyond belief. The theory does not allow the evolution of thousands of new species following catastrophes even if the purposed millions of years is taken into consideration.

Grand Canyon

The theory for the formation of the Grand Canyon was water erosion. The reason for such massive erosion is given that the surrounding plateau is at a high level which resulted in a very high flow rate for the river. This is the only instance of such massive erosion.

A more believable scenario is that the canyon was formed by a rift in the earth's crust. This can be justified by the close proximity of a mountain range. If billions of cubic yards of material were eroded away, where is the river's delta? Since the river flows into the Gulf of California, where there is no extensive delta, maybe the gulf is also part of the rift.

Great Lakes

Another instance of a theory that runs into trouble is that of the great lakes. It is theorized that the great lakes were formed by the recent ice sheet. It is difficult to imagine that three such large depressions can be caused in one location and no where else under the ice sheet. The depth of the lakes along with salt deposits underneath also does not support the ice sheet theory.

A more logical scenario is they were caused by a meteor from the north entering at a shallow angle impacting billions of years ago. It would have to be assumed that the meteor broke into three major pieces, each of which formed an elongated crater. The smaller lakes would have been formed by smaller pieces of the breakup. The salt deposits require that the area must have been connected to a sea in the past.

Heavy Metals

A current cosmological theory states that heavy metals on earth were attracted from the debris from exploding stars. This tends to indirectly suggest that these metals should be evenly distributed across the earth. This is not the case since the metals are found in local concentrations called veins. This suggest that the metals form during the planet formation process.

The concentration of hydrocarbons is attributed to the decaying of plant matter. Although the geologists do not refute this belief they know that the creation of hydrocarbons is the result of additional factors in a planets evolution.

Dinosaurs

The study of the dinosaurs has gleaned most of research efforts of the last century. The number of theories about their demise number in the thousands. Reasons purposed are droughts, big mosquitoes that spread disease, and a meteor strike. The meteor strike of which there is evidence is the most believable theory.

Theories abound of how species evolved across the catastrophe boundary into very divergent species which the theory of evolution does not allow. It is difficult to imagine how a creature as large as a dinosaur can shrivel to the size of a bird. It is just as difficult to imagine how the thousands of species of birds and animals evolved from the purposed process. Maybe the XTC liked feathers and installed the birds after the catastrophe.

The extinction of the mega fauna is attributed to over hunting by man. On must ask how did he achieve this with crude weapons. It's more reasonable to assume that they drowned during the great flood because they weren't on the boat. All the numerous species of birds and animals which did not exist with the mega fauna did not evolve from nothing and they were not on the boat either. The tarraformers installed them after the flood.

Ancient Meteorite

The following is the suspicion of an ancient meteoroid impact. This essay is included to the prevent the loss of information that may be of benefit to some future researcher.

In the early 1950s I was a young boy. I played with a rock that looked like an elongated brown potato. I realized the object was something special. So guess what, I used a hammer to smash it. What I found did not have any meaning at the time. I did not have any knowledge of meteorites.

I saw particles that I thought was gold but later learned was nickel. The broken surface soon developed a rusty color. After learning the accepted composition of meteorites I realized that this is what is classed as an iron/nickel meteorite. As time has passed I have come to realize the meteorite deviates from the iron/nickel class. The fragments contained a chalky inclusion the size of a small pea which does not fit the iron-nickel description.

My family was corn producers. In the middle of our cornfield there was what my parents called the "ole"pond. Although I never saw any standing water, it can be assumed that the depression was filled in by erosion of the surrounding crop land. In the early 1800s it did contain potable water as evidenced by a Creek Indian village nearby.

I have come to believe the depression is the result of a meteor strike in very ancient times. Facts to support this belief are listed as follows;

1. The presence of a meteorite in the general area.
2. The depression is more than thirty feet above the water table which rules out the possibility of a marsh.

3. The argument for an extreme age is the fact that there is no ejector surrounding the depression. The subsoil in the area of the depression is red clay, which if ejected would be incorporated into the sandy loam topsoil over time.

4. The depression is circular with a diameter of about one hundred feet.

This anomalous depression is not singular. About thirty miles south there is another referred to by the old timers as the "halfway pond". This pond contained potable water in the late 1800s, since it was used as a campsite for wool producers on the journey to the wool market. I never saw the halfway pond but learned of it from my uncle.

If one subscribes to the meteor strike belief, it can be extrapolated that both anomalies were caused by a meteor that broke up upon entry into the earth's atmosphere. It can also be theorized that a more massive impact may have happened off the Mississippi gulf coast. Since the size of the presumed impact area is significant it would be fairly easy to validate a possible impact with modern metal detectors.

10-You Decide

The investigation of any event requires the answer to the standard questions "How,Who, Why,When,What". Acceptance of the tarraforming theory will answer the question of who built the ancient constructions and who pilots the UFOs. All the other components of the queries will require new theories which will replace current theories. The new theories will remain guesses until the XTC decides to make contact.

Before we close this section I must share a rumor with you. The XTC determined that the XTB assigned the task of monitoring earthly TV broadcast were wasting time by playing fantasy football.

Since it is difficult for man to envision an intelligence that may be millions of years more advanced a new discipline will need to be formed to allow the interpretation of the fingerprints left by the gods

Section-5
CIVILIZATION

WHEN ONE HEARS THE word civilization he more than likely will think of ancient Greece and Rome. There are many more ancient civilizations but they did not have as great of influence on modern civilization as those two. They were instrumental in developing law, architecture, and engineering. Although the Romans utilized a very cumbersome number system it was still able to excel in engineering.

Following 356 BC a Macedonian general by the name Alexander proceeded to conquer most of the cities of the ancient world. He was very successful but died at a young age.

Following his death Greek civilization was absorbed into the rising Roman civilization. The roman empire was able to extend its influence to all land areas around the Mediterranean sea, and even as far away as England.

As time progressed it came more and more difficult to maintain its influence. This required many foreign wars to be conducted. This resulted in the evolution of a strong military which resulted in the creation of strong generals. The only path of advancement was to control the government, but to do so each general was required to contend with other generals with the same goals.

In 218 BC a Carthaginian general named Hannibal decided that his city would no longer submit to Rome's dominance. Although

Rome won the ensuing war it became aware of the problem of over extension. The conflict between generals finally manifest itself in the decision to separate into an Eastern and Western empire. This only served to accelerate the disintegration of the civilization. It was at this time that Christianity was made the state religion of the eastern empire.

2-China's Evolution

China is an instance of a very old advanced civilization that has become recognized as such only recently. It has had periods of brilliance followed by warfare between disparate tribes. Then the tribes were consolidated into a single government. This government remained isolated for centuries by choice. This isolation was shattered by the Japanese invasion during WWII.

China's written language tends to suggest to have been influenced by the XTC. The complexity of thousands of characters falls between the Egyptian hieroglyph and the Sumerian abstract.

China has gone through an extreme evolution in the last few decades. While the US and Russia was expending their resources competing with each other in the cold war, China was developing its form of government through evolution. The first attempt at a new form of government was the communist creation of farm communes. This attempt failed because there were no economic incentives. Even more fatal was the destruction of the knowledge base.

After the abject failure of the great leap forward experiment the communist party decided to allow an economic incentive for production. This meant that the knowledge base had to be reestablished and the rural population had to readjust to producing food for profit. It was realized that brain power could facilitate the production of goods that could be sold on world markets. In addition the rural population would support the cities with food. China's version of capitalism has worked so well that it has advanced to the number two economic power. In five years it will be number one.

The reason the new plan was instituted so quickly was the fact that the government did not have to contend with lawyers, special interest groups or environmentalist. The new plan also has its flaws. Now instead of everyone being poor there is the problem of the people of the cities being resented by the rural population because of their affluence. The following is a plan to allow those with pitchforks to share in the benefits of civilization.

3-China's Future

The three gorges dam was a tremendous achievement. It is guessed that all power produced is not utilized. If it is assumed that all the land around the dam within a hundred miles is not rock, then the following plan will be feasible.

Excess power can be used to pump water into a pipe line to provide irrigation. The intake of the pumps should be below the main dam to prevent the depletion of water used for power generation. This can be thought of as using the water twice. The pump intake may require a small dam or excavation to allow sufficient water supply to the pumps.

This plan does not have to rely entirely on the three gorges dam. Smaller dams can be constructed which have enough electrical power to supply a rural area with water and electricity. The small rural communities will be able to produce needed food.

The pipe line will be terminated in an area that can support farming. The terminus should have a collection pool. This pool will supply irrigation water on demand as well as water for household consumption. In addition, the pipe line right of way will include power transmission lines.

The terminus of the pipe line can support a small rural village. The residents of this village will supply food needed in the future. The farmers will have the incentive to produce because of their participation in the well being of the country and self worth. Self worth results from an economic incentive.

A geographic area that includes several of the terminus areas should have a food cannery.

This solves the seasonal availability of crops as well as transportation to the cities. As more of the population of the rural areas can afford to buy goods produced in the cities the country will have to rely less on exports.

4-Population Growth

The population of the world is increasing exponentially. This is due to more food and healthcare. The earth has reached its carrying capacity. An increased population requires more goods and services. The increased demand for goods results in a more increased demand for limited resources. An analogy with wildlife populations can be made. When a population exceeds the environment's food supply it generally dies off due to starvation and disease.

The first problem facing overpopulated countries is the addition of young people to the work force. They expect to have a job. If they cannot secure a job they will secure a pitchfork. The lack of jobs widens the spread between the haves and have nots, thereby increasing the size of the democratic voting block. This supports the philosophy of "take from the rich and give to the poor". This makes everyone poor with the poor even more poor. These most poor will be first to die off.

During the big economic downturn the president promised to create or save millions of jobs. Since the government can do little to create jobs in the private sector, it increased departmental budgets to allow the creation of new jobs and retention of existing government jobs. These jobs increased the deficit, but did little to boost the real economy because they did not increase the production of goods and services.

Admittedly the phony jobs did affect the unemployment rate. The unemployment rate is manipulated to appear more acceptable than is really the case. The rate does not include many sectors of the job market and those who have exhausted unemployment payments.

The reason for the lack of jobs for young people follows:

1. Productivity eliminates need for many jobs.
2. Cheap labor in other parts of the world eliminates need for local jobs.

3. Job seekers are not able to qualify for available jobs, although they have degrees.
4. The US creates fewer and fewer manufacturing jobs due to labor cost.
5. High employment cost of union workers forces fewer employees.
6. Population growth outstrips all of the above.

The second biggest problem facing overpopulated countries is food production and distribution. Most developed countries in arable locations have developed food production and distribution systems. This has led to what is known as mega-farms. Mega-farms foster the system of canneries and transportation systems. This insures that non-farm segments of the population can acquire needed food. If the non-farm segment ceases to function the farm segment will also cease to function and visa versus. That is to say that the two segments are interdependent.

If the food segments are interrupted the population problem will revert to one of malnutrition. The only people that will have food are the rural population which can grow enough food for its needs. The rural population will have to resort to home canning to compensate for the seasonality of crops. Any excess food of small farms can be inefficiently distributed to the cities. No number of small farms can ever replace the production and distribution of mega farms.

The disruption of the food supply will result in anarchy and disease. The cities will be hardest hit. The result will be detrimental to civilization. The knowledge base is in the cities. The only remnants of civilization will be written documents because modern forms of storage will be inoperable. After a few generations the written documents will deteriorate to the point of no use. The preservers of knowledge will not be able to trade their efforts for food.

5-Epa's Dangerious Agenda

I was going to start this paragraph by stating the EPA was misguided, but then I realized that would be erroneous, since its reason for existence is determined by special interest groups. In addition government departments can enlist the EPA's assistance in developing pet projects in return for submitting to its demands for its pet projects. When you realize that your politicians are subservant to both special interest groups and the EPA you have a real stinking kettle of fish.

The CO_2 produced by transportation will be drastically decreased because of the decrease in population and the need to travel to a nonexistent job. An effort currently underway by the EPA to reduce CO_2 levels by regulation will have had no effect. It is penalizing power generation companies with ridiculous regulations in hope of driving up the price of electric power. It is also penalizing the producers of hydrocarbon fuels to drive up the price so that alternate energy will appear feasible. China will love this because it will create a market for their alternate energy products. It might be wise for one to invest in candles and bicycles before the collapse.

The greenies will love this in the short term because a collapse can do more to restrict CO_2 release than anything that politicians and special interest groups can do. It is a scientific fact that the earth's climate changes periodically and man can do nothing to to affect the change.

Congress created the EPA and did not realize they had given it the power to issue mandates of which congress had no control. Therefore, congress keeps quite or in some instances uses the mandate to advantage. A future congress can only defund the EPA and rename it APE. The following lists the areas where the EPA's actions are most detrimental to civilization.

1. Climate change. It is going to happen and nothing man can do will affect the outcome other than adapt to the change. In an effort to stop the unstoppable lets destroy the economy trying.

2. Extinct species act. Let's penalize those mean old farmers and divert their water to protect a species that no one ever sees. Lets protect those desert dwellers by preventing transmission lines and pipelines. These

efforts can decrease the human population to the point of itself of becoming extinct.

3. Clean air and water act. This act can be applied to almost any human endeavor.

Each of these areas spawn many supporting mandates. Mandates need not be associated with any particular act; that is to say a mandate can be attached to any act or existing mandate.

The greenies coined the term global warming because of the release of CO_2 which was claimed to be man's fault. All this concern came because of the observation of the melting of the ice caps. Then it was learned that the ice caps have been melting for thousands of years. After the greenies became aware of the scientific knowledge of the earth's cyclical climate change they decided to use the term climate change to increase the possibility of being 100 percent correct in their predictions. The greenies have not been able to correlate CO_2 with earth quakes, but they are working on it. "joke'

As time progresses as it always does more and more of the knowledge of civilization is relegated to electronic media. If the electrical driver of this media is interrupted, then the civilization would have been assumed to have no knowledge of writing. The only method of preserving the knowledge of writing is books. Books will survive only a few hundred years without special care. Religious archives were paramount in the survival of ancient text. Much ancient text was preserved by being chiseled into stone. In modern times we no longer emboss stones but plastic is embossed which can survive for at least a thousand years in a fragile state.

The US democracy is so convoluted that special interest groups can encourage politicians to pass legislation that support their cause. Much of the legislation is detrimental to the survival of the county. When the US civilization collapses many others will collapse.

6-A Tale Of Hard Times

During the great depression the rural population supplemented the un-seasonal portion of their diet with wild life such as deer, rabbits, and feral hogs. My family was blessed with feral hogs and even wild chickens. We marked our hogs which is analogous to branding cattle. It involved the notching of an ear for identification. After the depression subsided the poorer segments of the population continued to utilize the feral hog as food. To them the marks did not enter into the selection process and owners of the marks no long cared.

The following is an illustration that there are still morally good people even bad times. In the early fifties a group of Mennonite religious members moved into Mississippi. Our farm was adjacent to the farm of a Mennonite. The Mennonites were very productive neighbors After a few years the Mennonites decided to go back to Ohio. Many years later my father received a check for twenty dollars from our neighbor. The explanation was that the neighbor realized he was dying and he wanted to atone for his sin. What he considered a sin was that he killed one of dad's feral hogs that invaded his corn field. This was not what he considered a sin, but the fact that he used the remains as food was a sin.

7-Prophesies

Prophecy predicts events that are believed to happen in the future, but no specific date is specified. When a particular event occurs existing prophesies are searched to determine, if the event was included in the prophecy. Since prophesies are vague any match is questionable. Prophesies are generally based on events that has happened in the past. These include natural catastrophes and political upheavals. The big one is the end of the world. This one is plausible since both religion and science has shown that it has happened in the past. These catastrophes affected more than civilization but life itself.

One natural catastrophe that could be disastrous is the large eruption of a volcano where the sunlight is diminished by particles in the atmosphere. This could be the answer to the bottlenecks in the human population found by DNA researchers. Within the bottlenecks the human population could have been as low as three thousand world wide.

The only event that can be predicted far into the future is one of astronomical nature. Modern scientists have the ability to predict orbital paths of objects that will occupy a point coincidental with the earth in its orbit far into the future. This requires the object must have been observed so as to compute its orbit. When an orbit is computed the exact time and date of the point of collision can be predicted.

The problem arises when the object has such a long period that it has not been detected and monitored in modern times. The Sumerians documented the existence of a planet that had an elongated out of plane orbit with a period of 3600 years. They referred to the planet as the Planet Of Crossing (POC).

One of the POCs close approach may have to provided the reason for the Mayans to start their calender in 3114 BC and Jews to start their calender in 3760 BC. The Mayan calender has begun to be used in prophesies because it clearly specifies an end date. Many attribute the specified end to the depletion of space on the disk. When one realizes the the ending date is one of the most important of the calender the law of averages is against such ending to occur by chance. This requires the carvers to supply a specific end must start at the end a carve in a reverse direction. This requires the question to be ask; did the calendar begin in 3114 or at an earlier date not displayed for lack of space.

Since the Mayan calender ends on a specific significant date it is assumed to contain a prophecy. The only event that could be known far into the future by its creators is of astronomical nature. The event must to have happened in the past to allow data to predict its re-occurrence. This implies the previous occurrence did not destroy the earth because man continued to exist to mark the event. The existence of the Mayan calender suggest that the XTC transmitted the knowledge to man since man was very primitive and could do only what they were told.

8-War's Impediment To Civilization
Since the US has lost all the significant wars since WWII you would think the military would educate themselves as to the cost

of the outcome of any action. It wastes it resources an life without a knowledge of an outcome. Russia bankrupted itself fighting a propaganda war with the US. The US has been wasting its resources in wars intended to promote democracy in countries that cannot support democracy. Even after the war is concluded resources must be spent to replace the infrastructure destroyed and establish a resident maintenance force. This leads to the philosophy of nation building to encourage the hoped for democracy. Even after all the expenditures many a democracy does not survive.

In the future, if the US continues to pursue the idea of democracy, the infrastructure of a country should be destroyed by air and imports eliminated by sea. This does not have the cost of boots on the ground. After a few years the population will have suffered sufficiently to accept democracy. How long it will last is unknown.

Section-6
ENVIROMENT

ENVIRONMENTALISTS DEVELOPED THE THEORY that man is causing global warming by his creating CO2. The theory was given a boost by former vice president Al Gore when he released his book "EARTH IN BALANCE" after his loss for the bid for the presidency. He made a lot more money supporting the theory than he would have made as president. Scientist not supporting the theory showed that the data used to support it had been fabricated to insure a heightened fear factor.

After it was seen that the theory did not meet the scientific data the name of the theory was changed to climate change. This would insure the theory would have a 100 percent chance of being right. Science has proved that the climate has changed many times in the past. Most of the changes were in cooling which resulted from gasses emitted from volcanoes. These gases are still being emitted today but at a lesser rate. These gases were not emphasized in the global warming theory; instead man was accused of producing too much CO2.

The single most used event to support the theory is the melting of the polar ice caps. What is not stated is that in the past most of the northern hemisphere was under an ice cap. It has been melting for thousand of years.

The EPA has taken control of the climate change theory to justify the reason for its existence. If man is causing the problem then one must conclude that over population is the root cause of the pollution. The EPA can do nothing to directly control the pollution problem, but it can destroy the economy which will decrease the population. The first step is to destroy the food supply and distribution networks. After that everything else falls into place.

2-California

An example of extreme environmental laws passed in an effort to control CO_2 happens in California. This is the result of its most populous city having an air pollution problem. The problem stems from the fact that the city is located in a geographic area defined by natural restrictions. Since the state was very wealthy at the time it was decided to tax unaffected areas to correct the problem? All the efforts have failed and the result was to decrease the state's wealth many fold. California will no longer be considered the golden state but the state to be avoided.

The latest extreme mandate has been announced that requires electricity generation by alternative sources will be thirty percent by 2020. The inefficiency of alternate energy does not scientifically allow the required capacity. The edict does not address the CO_2 generated in the restricted area of Los Angles generated by transportation. After a period of subsides the electrical companies will begin charging realistic rates which will cause more tax paying companies to relocate thus reducing the tax take.

California has more special interest groups than any other state. When you scrutinize the membership of these groups you will find that very few have sincere concerns about their cause. Once they get the ball rolling a special interest group forms whose members can then draw large salaries. Special interest groups have begun using the extinct species act to limit the production of food. This is the first step in reducing the population.

3-Ethanol Boon-Doggle

A boondoggle initiated by the EPA to combat CO_2 was ethanol to replace gasoline. This impacted the food supply. Ethanol is less efficient and more costly than gasoline. It is costly to transport ethanol from Iowa to California where it is assumed most needed.

The project was based on an erroneous model. It was based on Brazil's self sufficiency in transportation fuels. The fallacies in the model as it applies to the US follows.

1. Brazil had a small transportation fleet.
2. It had a large agrarian work force.
3. Its primary crop was sugar cane.
4. The process of extracting ethanol from corn is much less efficient than from sugar cane.
5. Ethanol cannot be transported via pipeline because of corrosion.
6. The US diverted its workforce and resources from producing food to producing ethanol.

Although the EPA has the power to destroy the US economy it has no power to affect other countries. As other countries become more advanced their transportation fleet will increase in size. This will negate anything the US does to limit CO2. Very few countries will risk destroying their economy to support the theory. These countries don't have special interest groups controlling their politicians. The US will push for punitive measures against countries it deems to be polluters. At that time we will not be able to throw our weight around because we will no longer be a world power.

Science has verified that climate change can occur and man can do nothing to prevent its occurrence. Climate change will solve the problem of over use of resources because the new population will require less. Climate change will set the evolution of man back thousand of years because the knowledge base is in cities which will be hardest hit. Rural populations will subsist if the change is not too great because they can produce food.

4-Survival Of Fittest

The following illustrates the efforts that the EPA and its special interest groups will initiate to justify their existence.

A few months ago a couplet of commercials commenced airing on most cable channels. The commercials were aired as a public service. The first of the couplet fits the definition of a public service. It targets children with instructions for applying CPR to heart attack victims. The instructions specify calling 911 and then pressing on the chest until help arrives.

Anyone that has watched the playback of a 911 call notices that it takes three to five minuets to supply the information required by the 911 operator. By the time the 911 call is completed it will be too late to apply CPR. It would be more effective to apply CPR then call 911. There are very few children who watch the channels that are airing the commercials.

Sometime later the airing of the first member of the couplet has been suspended since its purpose of acclimating the viewer to the message of the second member of the couplet. This second commercial retreads the scare tactics used in previous commercials about the dangers of lead paint.

The statement is made that one million children are affected by lead paint in homes built before 1978. What was not mentioned is the fact that the seller of any one of the many millions of houses built before 1978 must pay for a lead paint inspection and pay for remediation if lead paint is found. The inspection applies to the exterior of the house where children have no access. In most cases it is more cost effective to demolish the property than pay for the remediation.

It is hard to imagine that a special interest group such as leadfreekids.org would have the resources to sponsor the airing of the commercial at such an unbelievable rate. It is more than likely that the EPA created the site to hide their sponsorship.

The following is a list of other lead paint scare tactics sponsored by the EPA.

1. The decline of the roman empire is attributed to the use of lead in the piping of the water system.
2. The demise of a polar exploration party is attributed to the lead solder used in the sealing of food tins.

3. Lead in solder threatens a home water system.
4. Lead in gasoline was polluting the environment.
5. Children of impoverished families were eating lead paint flakes.
6. Chinese toys were endangering children by the use of lead paint.
7. The EPA realizes that it can get more attention if they use the threat to children as a primary scare tactic.
8. As this essay was being composed a news flash stated that environmentally safe shopping bags made in China were found to contain lead.

It can be seen from the preceding list that the EPA has lots of arrows in its quiver to justify its continued existence. The best way to limit the EPA's adverse effect on the economy is to de-fund it and limit the punitive lawsuits it can bring. A good name for the de-funding effort could called APE, EPA spelled backwards.

After this essay was initially composed the original member of the couplet has been replaced with another public service commercial which shows the US has fallen behind in education. What is peculiar is that the lead paint is aired before and after the new addition. One has to ask the question do the TV channels have so little demand for advertising that they have to resort to the cut rate advertising to fill a time slot. Another possibility is they wish to evade the ire of the government by not airing the propaganda.

The use of asbestos which was banned many years ago still supply a livelihood to lawyers who seek access to the trust fund set up for remediation many years ago.

5-Electrical Boondoggle

The current replacement for the failed ethanol project is to subsidize the development of electric vehicles. This project is already showing its faults as the following list shows.

1. High initial cost.
2. Limited mileage per charge.

3. Even more limited mileage if heating and cooling is used.
4. Long charge times compared to gasoline fill up.
5. Short battery life.
6. Very high cost of battery replacement.
7. Unreliable source of lithium used in batteries. The most abundant source of lithium is the undeveloped country of Bolivia which only has access to the sea via Chile or Peru. Most battery technology stems from China.
8. Questionable resale value.
9. Electric utilities will not be unable to meet increased demand due to restrictions placed on them by the EPA. The price of a charge will increase accordingly.

One solution to the mileage problem is the development of the hybrid vehicle. A hybrid vehicle does not solve the problems of a dead battery and battery replacement cost. It does eliminate the need for tow trucks when the discharge limit is reached.

Solar Energy

The production of electricity by solar cells is so intermittent and inefficient that it will never be a viable solution for generation. It can possibly fill some niches. One such niche is charge stations. This can be achieved by covering parking spaces with solar panels. The question to be answered is how well will solar panels withstand hail storms? Another niche is to use solar power to generate hydrogen. This eliminates the cost of battery storage and utilizes the less costly portable tank storage. There have already been attempts to develop fuel cell powered vehicles which failed because of the expense of hydrogen. Hydrogen generated by the use of solar energy can change this.

Natural Gas

The only possible solution to the problems in the short term is the use of natural gas for transportation fuel. Natural gas is efficient and emits less CO_2 than gasoline. The world is wasting vast amounts of natural gas in oil fields where the gas is flared as a waste product.

The US has vast reserves of natural gas. The increase in the supply that may be needed could be vastly increased, if the greenies are removed from the decision making process of DOT and EPA. The use of natural gas will not require much new technology. In fact the technology for gas powered engines exist today.

6-Transportation Cost

Many of the costs of transportation are not recognized by the driving public. This cost is spread out over many areas which can be viewed as a tax on consumption. The best way to analyze the many areas of cost is the list form.

1. Tax and title at purchase time.
2. The DMV is one's first interface with the transportation department (DOT).
3. The DMV is mostly state controlled. The intelligence quotient of most DMVworkers is very low. This is likely the result of quota mandates issued by thefederal government.
4. License plate fees supply money to build and maintain highways. This money is normanally diverted to other purposes.
5. Environmental inspection fees are just another way of collecting taxes. Most of the criteria are specified by the EPA.
6. Gasoline tax is supposed to be used for highway maintenance. It can really be used for any thing the federal government chooses. Even to subsidize alternative energy projects which will never be cost effective until all sources of fossil fuels are depleted. At this time the population problem will have subsided and demand will have decreased.
7. Seat belts have been proven to save lives in the event of a crash. There is no identifiable cost unless you are found not using them.

8. Insurance has become mandatory in many states. This eliminated the need for uninsured motorist insurance. Some states have the ability to determine that one has a driver's license, but does not have insurance. That gives them the ability to jiggle one's chain even though he does not drive.

9. Many municipalities lower their speed limits to create a speed trap to allow ticket generation.

10. The mandates listed above give local enforcement agencies an easy way to meet payroll by writing tickets for non compliance. If none of the above apply, then the old standbys of faulty lighting and speeding can be used.

11. The newly designed cost associated with electric vehicles is not yet realized. This cost includes high initial cost, high electric bills and the big one is periodic cost of battery replacement. The life span of a combustion engine is determine by the time in operation, while the life of a battery is determined by time in existence.

Section-7
INVESTORS

DURING THE PAST DECADE US economist bragged about a consumer based economy. This was an effort to deflect criticism for the loss of manufacturing jobs. Job creation is at a level that does not keep pace with population growth. Even those jobs associated with the selling of goods of other countries are not enough to close the gap. If young people cannot find jobs you have real problems.

The performance of any economy depends on the reinvestment of profits. The investments of profits are determined by the risk of doing so. Over the past few years this risk has increased to the point to where investments are not being made. The risk is becoming greater due to the government's philosophy of forcing the rich to pay their fair share. Since over half the population does not pay taxes, this amounts to confiscation of wealth and giving it to the poor. Don't forget that the government gets its fair share first. The reasons for taking from the rich and giving to the poor are jealously, moral, votes and preventing anarchy.

2-STOCK MARKET

Most laymen view the stock market indexes as in indication of the condition of the economy. These indexes are misleading because of the advent of program trading. This involves the trading of large amount of funds over a very short period of time. If a trader misses a minuscule gain he only has to wait a few days for the opportunity to make the same trade again. Another area where vast sums slosh around is that of short selling. In many cases this can determine future price of a stock.

Before the advent of program trading the mantra was buy and hold which no longer works. Using this philosophy one can invest in stock and wait a number of years for it to appreciate. At some point in time the market takes a tumble. The reasons a for tumble are many fold. The result is that gains made over the years can evaporate within a few days. The sad realization is that it may take many years to get back even. The inevitability of these cycles dictate that when you have realized a targeted appreciation sell the stock. You can buy the stock back later at a lower price or buy another with more potential.

Regardless of the ups or downs of the market the performance of the economy has the deciding effect. If companies cannot secure funding, their stock price will decrease. The short sellers will become overactive for a short time. If this happens to enough companies, the amount of funds deployed by the program traders and short traders will be diminished to the point that the market tumbles.

3-Physical Gold

As more and more of the possessors of paper money detect its limited survivability the funds reinvested will diminish. Investment in US treasuries although there is no return is considered to be the last method of preservation of wealth. This is coming to an end. Physical assets will receive more investment. Ownership of physical gold will receive the major portion of investments. This is because it is realized that gold is forever and governments have difficulties controlling and taxing ownership.

Although investing in gold enhances the investor's wealth it does not contribute to the economy. It does not pay taxes and contribute to investments intended to recognize a short term gain. Depending on how well you hide your hoard even your heirs will even find it hard to inherit it, but they will not have to pay inheritance tax.

One's investment advisers will be vehemently opposed to gold ownership since it will threaten their employment. When the elimination of salaries, capitol gains tax and inheritance tax is taken into consideration gold ownership is competitive. When the possibility of economic collapse is considered the ownership of gold wins out. Over the past decades the appreciation of gold has outpaced paced paper assets many fold. The downside to gold ownership is when it needs to be converted back into a currency. This will involve taxes that may not even be in existence today. In the event of a collapse the conversion to a currency will not be a problem because the gold will become a medium of exchange.

4-Gold Bugs

Now we need to analyze the thinking of "gold bugs". They have always been with us in one degree or another. The ownership of gold is driven by the following fears. They are economic collapse, inflation, government confiscation and laws that prohibit ownership.

In the early 1930s the US went off the gold standard. Up to this time gold was to bulky for large transactions so paper promissory notes which represented a given amount of physical gold or silver was in circulation. After the new law was passed the promise of gold or silver was removed from the face of the note. The change went unnoticed thus allowing the government to retain the ownership of the metals. Coins were replaced with what is called cladding. This is the process of using a cheap metal blank enclosed in a thin veneer of precious metal. The government went so far as to make it undesirable to own gold. The elimination of physical metals facilitated the printing of notes as needed. This allowed the value of a note to fluctuate wildly depending on the stability of issuer.

Judging from the number of gold commercials on TV and the shakiness of the world economy it can be assumed that the gold bugs are becoming more active. Gold bugs will become more active when information shows that an economic collapse is pending. It is an historical fact that all empires come to an end and if you have gold you or your heirs will be well position for events that may occur.

Even if a collapse does not occur the normal appreciation will make the investment worth while.

Gold bugs occupy two segments of the population. These segments include "big guys and little guys". The little guys invest in gold for a peace of mind in the event of an economic collapse. They will be better off than those who invest in paper. The big guys have more to invest and they also have more to loose in the event of a collapse or confiscation. When the economic collapse occurs the price that was paid for physical gold will no longer compute. The dollar will be worthless and the owner of gold will be well off in an emerging monetary system.

The argument against owning gold is the storage fees. This ignores the fact that paper investments also have fees and are more risky. Reference GM bonds. Many of the bonds coming to market today may not be able to pay the interest. If an entity defaults on the principal it does not have to worry about the interest.

One would like to think that the government could not confiscate earned wealth while you are alive. This is not true. The IRS can claim that you owe an exorbitant amount in back taxes. If it is not paid then an exurbanite interest and penalties are added onto the principal. Then if neither is paid in time the IRS can confiscate any property with a value sufficient to pay the claim. Any excess profits will be transferred to special interest who participates in auctions of such property.

An instance of government confiscation just came across the news wire. An individual submitted a cache of old gold coins to the government for evaluation. The government confiscated the coins. The reason given was they were probably stolen. This will more than likely be justified because they were not turned in when the US went off the gold standard. The coins are estimated to be valued at eight million dollars. This amount will meet the department's payroll for quite a few months, ha, ha.

Anyone in economic power must realize he will eventually die. One of his most important decisions is how to leave his wealth to his heirs. Your financial advisers have been successful in increasing your wealth, even in spite of economic downturns. Then when you die the social vigilantes will be there to confiscate half your wealth. This percentage will probably be increased to seventy percent in the future. Using this fact it can be seen that if you invest in gold today and it goes down by fifty percent you will still break even.

It may be decided that it would be easier to have an offshore bank account. This is proving to be a bad choice, since the US is strong arming such banks to divulge the names of holders of such accounts. The holders of these accounts will be deemed to be bad guys by the social vigilantes. Although the IRS has not been able to attach these accounts they may confiscate US property of equal or greater value than any gains. This goes to show no matter how hard one tries to eliminate a paper trail the government has the power to recreate a trail when needed.

Keeping physical gold in rented vaults is a bad choice. Number one it leaves a paper trail. Number two, if the economy collapses the vault companies will also collapse, but they have your gold. The time worn statement "possession is ninety percent of the law" comes into play.

If one buys gold and takes possession of such, the paper trail ends there. Whatever is done thereafter can reestablish the paper trail. One must devise a clever scheme of hiding the gold that does not involve a second party. If the government survives it will be able to tax any transactions in gold. This could lead to the ordinary day to day transactions in which the government is ignored. The only option available to the government is to make it illegal to own gold. This will not work because governments come and go, while gold is forever.

When countries sell their gold hoards the price of gold will go down. This will be temporary because the gold bugs will see this as a buying opportunity thereby reducing the supply which in turn causes the price to go up. When a country sells its gold it can be assumed that it is in finical trouble, although the more reason to buy gold. Stock traders use gold to stash excess funds. If they all withdraw their funds at the same time the price of gold goes down. A recent example is when margin rates were changed.

It is well to realize that when you die anyone with knowledge of any hidden gold becomes a treasure hunter or government witness. This dictates that as few people as possible know about the hiding places and that they die before you do. Early death of an associate is no guarantee that the knowledge of the event is not disseminated. Rumors of gold are more powerful to treasure hunters than any actual evidence.

When you realize you are going to die you must direct your heirs to their hidden gold.

They must then take responsibility of their inheritance. Do not mention such in a will since this will create a paper trail. If you create a paper trail whatever is left of your wealth will go through the normal path of inheritance and tax collection.

No one can tell you how to hide your gold. This is an individual endeavor and if done correctly can land you in the company of those who "carried it to the grave".

5-Gold Producers

Gold producers can facilitate the ownership of physical gold. The producers could produce coin sized blanks, the reverse side of which is stamped in the shape of an eight sliced pie. Such blanks can be referred to as a GOLPIE. The front side will have a finished suitable for another company to add its identification. This eliminates the need for fractional ounce coins which creates a handling problem and increased purchase cost.

The size of existing coins will have a monetary value much greater than needed for everyday transactions. The owner of a GOLPIE needs only to use a cleaver and mallet to divide his coin into eight segments. The size of the segments will not be questioned, since it is determined by a standard stamping. Any measure less than an eighth of an ounce is unmanageable. With the acceptance of the GOLPIE the "pieces of eight" term of measure will once again become prominent.

For those that have existing coins but cannot use them because of high monetary value, there can be a GOLPIE stamp which will mark existing coins in eight segments. Then the clever and mallet can be used to achieve a usable size. At this time the pretty emblems on the coins will have no value.

6-A Treasure Hunter Tale

To emphasize the benefits of owning gold I must tell you a treasure hunter's story. This story revolves around the civil war. There was a group of outlaws that specialized in stealing slaves, horses, and anything else they could get their hands on. After the war they started robbing the populace of jewelry and coins. As time went on they amassed so much booty that they could not transport it. At this time they began hiding their spoils.

This is also the same time journalist began to collect information on their operations. The first book to appear was the leader's assertion that his mother drove him to a life of crime.

The second book was about where the booty was hidden. This book had a short life because the treasure hunters went wild. This resulted in the government forcing the end to the publishing of the book THE CLAN.

Over a period of time Jim Co plan's clan was killed of by authorities. He was captured and sentenced to hang. It is said that he sold his body for a pint of whiskey. As he was being transported to the gallows he said to the people running along side of the buck board "where are you going in such a hurry, nothing is going to happen till ole Jim gets there."

My uncle who claimed to have had a copy of the book said he loaned it to a neighbor, needless to say he was never able to retrieve it. The book spoke of hiding two kegs and one barrel of booty. A keg was a small wooden barrel primarily used to transport such goods as nails, therefore the name nail keg.

A farmer in hiding observed a buck board with a barrel and two black persons being directed by two outriders. A couple of days later the outriders passed by again, the black persons and buck board was not present which can be thought of as elimination of evidence. It can be assumed the farmer with assistance figured out what was transpiring and retrieved contents of the barrel. This can be deduced from the fact that some citizens of the area became wealthy in a short period of time.

I know this tale is true because many years later I saw the barrel and excavation. The barrel was lined with tar which showed the impressions of coins and jewelry two thirds full. One would have to be pretty dumb to assume tar sealant was needed to protect gold.

Even into modern times a construction company uncovered a nail keg of gold. The other keg of gold has not been found. It is said to have been thrown into a deep pool in a near by creek in an effort to elude a posse.

This emphasizes the point that even if all immediate knowledge of a hiding place is eliminated there are other ways that treasure hunters can develop a trail.

Section-8
MAN'S DEMISE

This section deals with events that could destroy the US economy/civilization along with the rest of the developed world. Most of the events are man made, but some are natural. The natural occurrences tend aggravate the functioning of an economy that is unstable. Many suppositions about man's demise include the end of the Mayan calender which implies a natural event. Many ancient prophesies predict man's demise but they can be attributed to social events of their time. In addition none specify a particular date. These prophesies are based on current expectations based on the actions of the government.

The phenomenal increase in population has brought the social consideration to the forefront in democracies where votes are determined. Government will tend to appease the socialist. This is a decision that has been proven wrong in the past. This proved to be the final stage of a civilizations collapse. Even after the collapse the producers will excel while the non-producers will remain non-producers, but they have no one to blame.

2-US DEMOCRACY

Many countries have adopted democracy willingly while others through force by the US. What the US has not come to understand is that if a country has a population with an intelligence quotient that does not allow reading and writing and has a tribal mentality it cannot support democracy. But the US spends much of its resources in an effort trying to institute democracy in those countries.

The US democracy plays the haves against the have-nots. In reality this could be viewed as a tribal mentality. The Republican Party tends to be supported by capitalism while the Democratic Party tends to be supported by socialism. The major supporters of the democrats are from labor unions, special interest groups and have-nots each of which can be thought of as a tribal entity. Major support for the republicans is from the wealth producers that could be thought of as a kingship. You guessed it, if have-nots outnumber the haves, democratic voters increase. If you tax the producers and give to the non- producers everyone will be non producers.

Upon rigorous analysis it can be seen that a democracy resembles a tribal construct in many instances. In a democracy the special interest groups gain control of the system. These groups control the chieftain (politician) against the wishes of the influential segments of the population. In a tribal system opposing groups will be eliminated immediately. The demise of a democracy can take many years as evidenced by ancient Greece and Rome. The following decades result in anarchy followed by dark ages.

3-Democracy Vs Communism

Following WWII the USA and USSR became locked in an ideological battle that became known as the cold war. Both countries developed nuclear weapons at a rapid pace. It was soon realized the countries could destroy each other. This lead to the coinage of the term mutually assured destruction (MAD). Even if there had been a nuclear exchange the law of large numbers would have come into play. The large cities would have been destroyed but the remainder of the population would survive. The remaining population will have to contend with nuclear radiation and lose of knowledge required for an advanced society. When Russia saw that it could not continue to compete against the US economically the ruler Khrushchev was ask what he was going to do about the US. His answer was that he was not worried about the US because it was going to destroy itself from within. It looks like his prediction is going to come true. It is not difficult to believe that Russia and China are pleased by the problems of the west.

The difference between the collapse of Russia and the US will be the resilience of the population. The population of Russia had little before the collapse which allowed it to adjust to slightly less after the collapse. The population of the US on the other hand has been very spoiled since WWII. Even the poor have done well, but this is changing because the poor wants more for doing less. This sentiment has lead to the demise of many civilizations in the past.

After Russia went bankrupt it installed a form of democracy of which the US did not approve. Due to Russia's size the US was not able to affect the outcome. The US continues to expend its resources on wars and bribes in order to install democratic governments. The US is hoping that the countries of the Middle East will adopt a democratic form of government. This is unlikely when it is realized that a major portion of the population is still camel drivers that cannot read or write. The only knowledge transfer is via the Islamic religion which is anti-civilization.

Following the cold war Russian nuclear scientist migrated to third world countries taking

their nuclear knowledge with them. These countries have developed their own technology which they will use to blackmail civilization. After the blackmail has run its course they will detonate a device in the country of their adversary. At this time the thousands of nuclear missiles held by the country under attack will be targeted against the offending country which will become a waste land for thousand of years. Even neighboring countries whether supportive or neutral will be affected since the devastation is not limited to national boundaries.

4-China's Gain

During the cold war China used the conflict between the US and Russia as an opportunity to modify its form of government. China tried to appease its poor population with the "great leap" program. They learned that the poor produced even less because of the government guarantee. After the death of the implementer of the great leap died, the very smart successors looked at the rest of the world and realized that pay for performance of capitalism had worked for thousand of years. The new government does not subscribe to democracy which allows for strict control of the economy. After a couple of decades this decision proved to be the right decision, since the Chinese economy has surpassed that of Russia and is second to the US.

The Chinese stayed clear of the ideological conflict between the US and Russia. Instead it utilized the time and resources to build its new civilization. You can be assured when the time requires the use of military force it has the resources to meet the requirement. The other rivals will not have the resources to participate effectively.

5-Fear Of Nuculear Power

Since the end of WWII we have had four nuclear catastrophes. The first was the result of war. The Japanese civilization did not cease to exist, but it did gain knowledge about nuclear fall out. Next came the Three Mile Island power plant incident. This never posed any threat but gave the special interest groups and news media good press. Then came the Chernobyl event. This turned into a disaster because no acknowledgment of a problem existed and no assistance was requested. The final event was the result of the Japanese earth quake. The resulting tsunami damaged a nuclear power reactor. The Japanese scientist seemed to have contained the problem in short order. The news media did flame the flames for a few days. The American nuclear scientists were miffed because they were not consulted. Why would the Japanese look to the Americans who caused the first problem "bomb" and has little knowledge of a modern nuclear power industry? They only have pseudo science supplied by special interest groups. The Japanese survived the first event so why not a lesser event.

The latest scare tactic propagated by pseudo scientist and the news media is that of global warming. It is a scientific fact that the earth's climate has changed many times in the past. The big blunder made by the pseudo scientist is trying to convince man that he is responsible for the change by his use of fossil fuels. The only feasible replacement for fossil fuels for power generation is nuclear power, which is opposed by fear mongers.

If one has the choice of supporting a theory that guarantees that he must go into extreme poverty to prevent an event that may occur after his lifetime, he will choose the here and now. However, the choice is not his to make. Special interest groups which can manipulate the government will control the agenda. Politicians are easy to control if allowed to participate in a popular movement. Many think that money controls elections, but in reality it only allows popular ideas to be turned into votes. The members of special interest groups have to have well paying jobs, so don't blame them.

If you do not wish to descend into poverty vote to DE-fund the EPA and rename it APE.

This of course is a joke because your vote does not count. Once a movement is started it requires time for it to run its course at which time the harm will have been done. The EPA will issue mandates to affect pollution. Remember that mandates do not require voter approval. All the mandates issued by the EPA tend to drive you further into poverty. If the global warming theory is based on man's use of fossil fuels any remediation proposed by the EPA is doomed to fail. The problem stems from that of over population. Any savings realized by mandates will be swamped by the increase in demand of an increasing population. The EPA does not address this problem because it does not "sell". Mother nature has ways of correcting over population problems

6-Prophecies

In ancient times priest would allow forward thinkers to propagate their expectations of future events if it served their ends. Many of these prophecies came to pass because of the fact that a foreign army was camped on their doorstep. In the cases where the immediate threat did not manifest itself the prophecy could be relegated to the future which could last for centuries with the support of religion. Current prophecies can be made via the news media which have a lifetime of only a few days. A prophecy stated in the print media has a longer survival time because of ease of future access and interpretation. This group of essays is no exception, that is to say you change with current data. This may result in a prophecy being dismissed altogether based on new data.

The Mayan calendar begins in 3114 BC and is relative young compared to the Jewish calendar which begins in 3760 BC. The Mayan calender has garnered most attention since it specifies an end date Many attribute the specified end to the depletion of space on the disk. When one realizes the the ending date is one of the most important of the calender the law of averages is against such ending to occur by chance. This requires the carvers to supply a specific end must start at the end a carve in a reverse direction. This requires the question to be ask; did the calendar begin in 3114 or at an earlier date not displayed for lack of space.

Since the Mayan calender ends on a specific significant date it is assumed to contain a prophecy. The only event that could be known far into the future by its creators is of astronomical nature. The event must have happened in the past to allow data to predict its re-occurrence. This implies the previous occurrence did not destroy the earth because man continued to exist to mark the event.

The bible speaks of the end of the world in Revelations. No specific date is implied, so it relies on the knowledge that major catastrophes happened in the past, so it is assumed they will happen in the future. The problem with revelations is that it implies that man is going to cause it because of his sins. The latest prophecy that man will cause the end of the world because of his sins is that of global warming.

A prophecy that originated during the cold war that the US and USSR would destroy each other and civilization did not come to pass. This result was not caused by the two antagonist agreeing to coexist, but one going economically bankrupt. Another prophecy at this time was that the increased sighting of UFOs indicated the XTC would not allow a nuclear war to occur. This prophecy was ridiculed by main stream scientist as directed by CIA's debunking program because of its fear of belief in the XTC.

The prophecy was based on the hope or fear of the unknown. At this time the tarraforming theory had not been proposed. It was not understood that tarraforming does not allow developing life forms to be interfered with. The XTC views the conflicts between populations of man the same as man views the conflicts between insect populations. If man succeeds in destroying himself the XTC will write the event off as a failed project and start a new project that does not include man.

7-POLITICAL PARTIES

The two political parties have the power to prevent the formation of a third party. Third parties can gain enough support to allow participation in one election cycle. Any of the party members elected is then absorbed into one of the existing parties. Once a politician is elected he is most times controlled by a party and not by the wish of the voters. If a voter is dissatisfied with the performance of a politician he can do nothing until the next voting cycle.

This emphasizes the joke "if you do not like your elected official vote him out next term". This does not take into consideration the harm the official can foster before the next election cycle. Although there is the process of recall, it is seldom used because the politicians needed to vote for it are reluctant because they feel if they set a precedent they may be next to be recalled. The US does not allow a vote of confidence recall.

At the time of this writing the Democratic party has gained control of the government and the voting population. This can be attributed to the following reason.

1. The economy commenced to collapse at the beginning of election cycle.
2. The expected response was let's kick out the republicans and give the democrats a chance.
3. Let's elect a black person in an effort to atone for our ancestor's support of slavery. No thought was given to the fact slavery has existed through out history and still exists today.

4. Tell the voter base that the rich must pay its fair share. This appeals to the fifty percent of the population that does not pay any taxes.
5. The other contributors to the voter base was special interest groups, unions, lawyers, and environmentalist. They all saw the Democratic party as a means achieve their goals.
6. None of the remedies tried have improved the economy. Regulations have counteracted any remedy to improve the economy.

8-Special Interest Groups

Special interest groups are prominent in popular movements in such areas as behavioral change, safety, environment, and economical. The aim of special interest groups is to get politicians to pass legislation to support their agenda. An interest group persuades a politician to satisfy its interest with support of his reelection with campaign contributions and increased visibility via the news media.

The most powerful special interest group is labor unions. The special interest of unions is to have legislation that supports the reason for their existence. Their interest is generally satisfied because they control politicians through massive campaign contributions. The environmental special interest groups are fragmented. but receive their clout from indirect control of the EPA. The EPA influence can be seen in many government departments.

9-Labor Unions

Labor unions were created in the early twentieth century to address the poor working conditions and safety of miners. As time went on the unions expanded into other areas. The latest expansion was to include government workers. This was a major achievement since these unions affect the lives of all non-union members.

Union wage demands make private industry uncompetitive with other areas of the world.

Companies in the northern US are even at a disadvantage with those in the south. Unions have a labor law that gives them the right to form unions in the south, but the employees reject the corruption of unions for the guarantee of a job. The rejection of unions is one of the prime factors used by foreign companies to choose the south as a location of plants. The quality of work is superior to that of unionized companies and wages are in line with the task performed.

A company cannot control the productivity or quality of work of unionized workers. A worker cannot be fired without the union's authorization. This is a counter to the philosophy of "getting rid of dead wood" or stated differently of a company ability to shed unneeded and unproductive workers.

Some of the tactics used by unions to get a company to bend to its wishes is to call in sick, work slow, and strike. An example of work slow occurred in an eastern city during a snow storm when the workers were told to raise the level of snow removal equipment so as to require multiple passes.

Much has been said about the union's right to collective bargaining. What is not stated is the union's ability to strike if their wishes are not met? This sounds like coercion.

Over time unions will control access to the few available jobs. New hires will be relegated to those who are the relations of union members. Access to government sponsored jobs is already mandated to go to unionized companies. The have nots will eventually mount their pitchforks against the unions rather than the government. Of course the government will have to aid the unions since it created them.

10-Public Service Unions

During the employment downturn of 2010 the government bragged about saving and creating jobs. These jobs were unneeded which could be classed as welfare and could have been an effort to politically control the unemployment numbers. The reasoning for the hiring and retention of the members of the following public service unions are expanded as follows.

1. More teachers in a failing education system may solve the problem. The emphasis placed on college degrees for jobs that do not exist and will not exist is misguided.
2. More fire fighters needed to protect property. The only protection afforded by firefighters is to adjacent structures. Generally when the statement is made that a fire is under control it can be assumed that all combustible material has been consumed.
3. More policeman needed to fight crime. This can cause the crime rate to appear to increase because what was not considered a crime is now classed as such to justify the additional funding.
4. Shore up the postal system which fewer and fewer people use.
5. There are many more government sponsored endeavors that do not produce tangible results that soak up the available tax dollars but they do create jobs. This eliminates the need to speak of welfare or the unemployed.

Much as been said about funding basic research to make the US more competitive. Even if a significant discovery is made it will not affect the cost of production enough to effectively compete with the cost of production of countries currently producing the product. Solar cells is a good example.

11-Environment Protection Agency

There has gotten to be so many departments it is hard to keep track of them. Some were formed because of special interest groups. Others were formed when politicians saw that they could gain political capital by supporting a popular movement. Congress may not budget enough to fit a departments supposed needs. In these cases the department will use its authority to threaten a company with an investigation. The investigation will be withdrawn if the company pays an arbitrary fine. This also makes it appear that the department is doing its stated task.

The EPA was created by congress and is controlled by special interest groups. Since none of the groups have a common concern about the harm done to the economy by statutes issued by the EPA, it can do immense harm to the economy. The EPA is one of the reasons for so many jobs being transferred overseas.

Statutes issued by the EPA can have very long lifetimes. Lead and asbestos in products was banded many years ago. Some many years later the statutes are still being applied. When asbestos was banned a remediation fund was established which has been a source of income for many lawyers. The EPA can be thought of as an octopus whose tentacles reach into all other departments.

The DOT along with EPA has chosen to ignore the possibility of using natural gas as a transportation fuel. Natural gas is an abundant US resource. It can be seen that the oil lobbyist would oppose such a consideration but it is difficult to understand why the greenies would oppose it since natural gas burns cleaner and more efficient than gasoline.

The Greene agenda is to support funding of uneconomical alternatives.

The DOE is subsidizing the wind and solar energy industries in an effort to replace the generation of electricity with fossil fuels. It has been determined that it will be many years before both combined can meet demand. Since neither can address peak loads, blackouts will become common. The biggest producer of environmental products is China who laughs all the way to the bank.

One of the latest departments being formed is that of consumer protection. This is an effort to cover up the government's mistake of forcing lending institutions to make loans without any consideration of the ability of repayment. Risk takers would secure a property with intentions of fixing it up for resale before the terms of the loan adjusted to reality. When the economy adjusted they were left holding mortgages they could not repay. The government wants you to believe that those old lenders were to blame.

The department may be more able to justify its existence if emphasis is placed on the problem of identity theft. This can cause a person's credit worthiness to be destroyed by unlawful acts. You would think that a government that subscribes to the rule of law would be concerned about these events. Victims of identity theft should be able to expect the government to protect them from unlawful acts.

12-INTERNAL REVENUE SERVICE

This is a personal horror story. My wife and I have kept the derivation our incomes separate. My wife received bonuses and stock options of which she did not pay taxes. I don't know if the issuing company notified the recipient of the tax liability or if she ignored it. You can bet the IRS was notified. The IRS does not question a tax filer immediately but files the information away for future use which can be as much as ten years. At the needed time the information is brought to the fore. The amount in arrears now includes interest and insidious penalties. Since the amount was so exorbitant a tax consultant was engaged to help in remediation. I don't how know effective he was. To meet the IRS claims I had to raid my 401k. I tried to secure a loan but was unsuccessful because I was over age. I went ahead with the required withdrawal which will make us appear rich in the next tax cycle. This will result in paying taxes on the money withdrawn to pay the tax penalty. This brings to mind the ancient quote "Render unto Caesar, Caesar's due" and we know what happened to Rome. To insure that you are not the focus of a similar incident be sure to pay the required taxes at the time realized or pay a lot more later.

To put tax collection into the proper perspective we must refer to ancient records. In ancient times the king's tax collectors would go into the countryside at harvest time. They would confiscate all the harvest. Any crops not harvested would be burned in the field. There was not any concern as to the producers survival. The only bright spot was that the king did not survive very long. Today's IRS has the same power of ancient kings to destroy life.

An illustration how powerful the IRS is can be seen in the convictions of mobsters that could not be convicted in normal courts. They were convicted of tax evasion on income from illegal income which eliminated the need to convict on alleged charges. The funny thing is that the mobsters never had a W2 form.

13-News Media

The news media is the most powerful force in the US democracy. It uses the constitution's guarantee of freedom of speech to its advantage. It is used by special interest groups to gain popular support for their goals. Once they gain public support it is easier for their lobbyist to interface with politicians. Politicians utilize the news media to air video clips in an effort to get name recognition during an election campaign.

The news media must develop content from any subject considered to be news worthy. They go so far as to "dig up dirt" on prominent persons and government. In doing so, they have the power to destroy any entity.

Recognizing the power of the news media one can see why a restrictive government keeps tight control over the news media. In many such countries the news media might be used to control the population by disseminating only approved information.

14-Lawyers

The government gave lawyers the power to sue for any reason. This power was fairly easy to secure since most politicians are lawyers that could not make it in their chosen career. The lawyers are using the ability to sue for anything too full advantage which is affecting the economy adversely.

The idea of class action lawsuits was developed to allow lawsuits using plaintiffs that were not affected by an action. The persons that sign on to these suits are either activist or misinformed individuals hoping to receive monetary gains. The monetary gains received as the result of a class action suit is minimal since the lawyer gets a major portion of any monies.

Lawsuits are prevalent among doctors and drug companies which drive up the cost of healthcare. The new healthcare law purports lowering healthcare cost but it does not address frivolous lawsuits. You guessed it; the lawyers were opposed to any exclusion.

The government created a feeding frenzy among lawyers over the recall by Toyota of some of its cars. This was done as an attempt to affect a foreign brand to enhance the survivability of its own band. Even after the government later found Toyota to be innocent of the charges the government is allowing the lawyers to continue their suits.

Another example of the government "throwing red meat" to the lawyers is the BP oil spill. This was supported by the special interest groups wanting to replace oil with green energy. Although BP promised to bare the cost of remediation the news continued to paint the company as foreign evil company. Even though government had very little knowledge about remediation of the problem they attempted to gain good press for making an effort. Their intervention served only to extend the problem.

The drilling ban that resulted from the problem was another win for the greenies. This action affected all oil companies, not just BP. In an effort to extract itself from the political impediments BP sold off many of its US holdings. It has since struck a deal with Russia to drill wells off the coast of the north slope on their side of the international boundary. The greenies lost this one.

Florida has opposed off shore drilling for years to protect its beaches. China has signed up to drill offshore of Cuba which is offshore of Florida. If China has an oil spill, the US will not be able to treat China the same as it did BP.

15-Environment

Although social legislation is harming the continuation of US society, environmental legislation is doing the most harm. This is the result of the global warming theory espoused by the EPA with aid of the news media. The EPA has been given the power to dictate the behavior of other departments as needed. It even has the power to influence its academic scientist to falsify their research. After the involved scientists were brought to task by their peers it was decided to rename the theory climate change thus insuring a one hundred percent chance of being correct.

The theory is based on the premise that man's use of fossil fuels is causing the problem. This ignores the scientific fact that the climate has changed many times in the past. No emphasis is placed on the fact that the many volcanoes around the world release much of the pollutants causing the problem. Most support for the theory is from the measured melting of the polar icecap. Thousand of years ago over half of the North American continent was covered with an icecap. It has been melting for thousand of years for some unknown reason. It is only recently that man has become sufficiently advanced to notice the melting therefore a reason had to be developed.

The EPA has not given thought to the fact that its mandates will do more harm than the perceived result of climate change. They will increase the cost of energy and transportation. This means that manufacturers cannot supply their plants with the required electricity and its workers cannot afford to transit to work. With such an outcome one must rely on bicycles, candles and hand cranks. Even with these extreme imposed sacrifices the historical record shows that the climate periodically changes and man cannot do anything to change that fact. If the climate is not changing the sacrifices will be touted as having corrected the problem. If the climate is changing you will sacrifice in the short term as well as the long term. Given a choice of trying to prevent what may happen within a generation or two or today's lifestyle vote for retaining the status quo.

Some of the most ominous EPA mandates can best be discussed in list form.

1. The climate change theory spawns many associated mandates. The theory dictates the use of all fossil fuels be eliminated. It is purposed that alternative energy sources be developed. This is to be done by regulation that makes the cost of fossil fuel production prohibitive. This drives up the cost of energy to the end user which will cause him   to accept the alternative energy solution. The problem is that alternative energy can be shown to contribute at most five percent of the required energy output within the next hundred years. Even this slight

projection depends on a significant reduction in the population. The fossil supply will have been destroyed long before this time which will lead to the decrease in population.

2. The failure of the ethanol mandate for transportation fuel is the first instance where an alternative is not viable. The mandate is based on Brazil's ethanol program. Brazil's ethanol program was a success because it had a small transportation fleet and it used sugar cane as the feed stock. Brazil is actively developing its fossil fuel resources. The EPA decided to use corn as a feed stock. Corn is much less efficient in production of ethanol than sugar cane and the mileage derived per gallon is less than gasoline. The US transportation fleet is so massive that even if all available land is used for corn a severe shortage would result. Even minor droughts could cause shortages. The unanticipated consequence of ethanol production is that it creates shortages in the food chain.

3. The latest mandate was to have cars achieve greater than fifty miles per gallon of gas. Since this exceeds the scientific capabilities of the combustion engine the idea of electric propulsion has been adapted. This solution has the problem of excessive electrical cost of recharging caused because of the penalties imposed on generation from fossil fuels. The cost of the batteries is evident in the initial cost of a vehicle but what is going to make a lot of people mad is when they have to replace a battery after a few years.

4. Solar energy has been proven to be uneconomical. This is due to the initial cost and the fact that sun does not shine for enough hours. The cost of storing excess energy for later use is exorbitant. The only possible application where it may prove to be feasible is as a cover over parking areas where the output can be used as charging stations for electric cars. The argument is that the price

will come down with wide use. The only problem is that the US has few facilities to manufacture the cells. But not to worry we can buy the cells from china.

5. Neither wind nor solar can produce a continuous source of electricity. Therefore the base load has to be supplied by natural gas or coal both of which the US has ample reserves. Restrictions on their production and use are the main thrust for the elimination of fossil fuels.

6. Natural gas is abundant, cheap and produces the least CO_2 emissions of any fossil fuel. It is unclear why the EPA did not mandate its use as a transitional transportation fuel. Maybe the natural gas producers do not have any effective lobbyist. You can bet that the many hundreds of environmental regulations were instigated by lobbyist. In fact the EPA can justify its existence only by being sub-servant to lobbyist.

7. Special interest groups who wish to prevent an animal or fish from going extinct uses the EPA to impede operations that is detrimental to their environment. In most cases this affects human food production. The members of these groups place the survival of a species with no contribution to man's existence above man's survival.

8. If man is really causing global warming then it can be assumed that there are too many men on earth. Mother Nature has ways of dealing with over population. She generally solves the problem by limiting the food supply which leads to disease.

9. The EPA will continue to drive the US into a more uncompetitive environment compared to other countries. The other countries will pretend to support the fight against global warming all the while producing the widgets the US will have to buy to implement its program. This is logical since the US has a minuscule manufacturing base and it is getting even smaller.

16-Education

Commencing in the early 1960s the government realized that the US was behind Russia in graduating students in math and engineering. It commenced a push to get students to major in math and engineering. What it did not take into consideration is the fact that everyone is not suited to such degrees. Many students changed their majors to biology and business management to which they were more suited. The number of engineering and math graduates was much less than hoped for. Guess what, the government is doing the same thing again, but this time it is not because of a military threat it is because of an economical threat.

The pre-college education of US students is falling behind the rest of the world. In the last decade the problem was recognized so the education department received massive budget increases. The budget increases were used to hire only a few qualified teachers, but mostly additional staff and salary increases. The salary increases came about with the formation of the teacher's union. Salaries always go up while the quality of work always go down with the advent of a union. This is understandable when the unions subscribe to the last in first out (LIFO) method of employee rotation. This eliminates younger efficient teachers in favor of older inefficient teachers.

Even the best teachers cannot perform if students have no will to learn. Homework is viewed as a punishment since the child cannot go home to watch TV or play video games. Parents are the most at fault since they see school as baby sitters while they are at work. Students with parents of Asian ancestry are instilled with the will to learn at home.

The only positive that can be attributed to the department of education is an effort to insure the children of Mexican Americans and other aliens learn to speak English regardless of their home life. This insures that a democracy can function more efficiently if it is composed of a population with a single language.

A bit of jealously can be detected when public schools find themselves compared to parochial schools. In fact the public school unions exert all kinds of damage control actions when such comparisons are made. Regardless of what the unions do, the parochial schools come out ahead. Another education format disparaged by public schools is the vocational schools. Probably the reason is that vocational schools train students for a specific job in two years. In addition the teachers are qualified and non union.

If one considers the time and money required to get a four year degree compared to its worth in securing a job he must question the benefit. Student loans are used to pay for the cost of acquiring a degree. The intentions are that the student will repay the loan when the student enters the workforce with a well paying job. Many degrees are not associated with any particular job and with the lack of any available jobs the student can not commence to repay the loan. At this time the parents must commence to repay the loan since they co-signed it. This can create severe hardships for families.

Most degrees do not prepare the holder for a particular job. A four year degree does serve the purpose of allowing the transition from childhood to adulthood. The department of education espouses the need for increased funding to educate for the jobs for the future. The argument against such funding is that even with the massive funding increases of the past many young people with degrees cannot qualify for the few available jobs. In addition in the future there will be fewer jobs that require degrees.

In time having a degree will be a less important guarantee in securing a job because of the reasons listed below.

1. Population growth outstrips job creation.
2. Productivity eliminates the need for many jobs.
3. Many degrees do not generate knowledge required for a job.
4. A degree is not required to work in a factory producing widgets. The problem manifest itself in that we no longer produce widgets. This is because the people that have degrees think it below their station to produce widgets.
5. Many jobs will be transferred offshore due to wage cost.
6. Few new manufacturing companies will be created.

7. Young people with a degree that cannot find a job are volatile.
8. Innovations by other countries will exceed that of the US.

The jobs available in the future will be selling goods produced by other countries. The only other possibility is a government job. The stimulus package was said to create or preserve jobs. This made the unemployment look better but it did not do anything productive. The jobs were not needed, and were nothing more than public welfare and a power grab by the government.

Following the great depression the lack of jobs although not caused by productive gains was designed to solve the lack of jobs problem. A program known as the WPA was instituted to create jobs. This was not a welfare program that allowed recipients to stay home and do nothing. It involved planting trees, digging holes and refilling holes. The participants could take pride in the fact that they had a job. Today one does not worry about securing a job but access to food stamps.

17-The End

The scientific method has advanced man's evolution more than any other event. The advancement is so significant as to cause a problem of over population. The US will be most affected by population growth. In the past it could produce enough food for its population. This will not be true in the future because of population increase. The EPA and other government departments controlled by special interest groups are making it so expensive to produce food that many farms are closing down. The US is unique in this restriction in food production. Other countries that have to sell resources to buy food will experience other problems. The die off will occur as the result of starvation, disease and political unrest. Over-population creates the following problems for democratic civilizations.

1. Resources are being depleted at an alarming rate.

2. The food supply can be endangered by misguided government regulation or severe and prolonged drought.
3. Non producers will out number producers.
4. In a democracy where the non producers out number the producers the government tends to favor the non producers because that is the source of the required votes. At this time the entire population becomes non producers.
5. Advanced countries must compete with less advanced countries on labor cost. Labor cost can be partially offset by productivity gains. Labor union's unrealistic wage demands which have support of politicians out stripe any savings made by productivity gains.
6. Productivity gains result in fewer jobs. This creates a problem for government because young people entering the job force do not have access to a job.
7. The lack of jobs can be magnified when a developed country transfers its manufacturing jobs to a less developed country because of a labor cost differential and regulations. This results in the only available jobs being for salesmen for the goods manufactured by those countries.
8. In the previous decade leading economist were touting the fact the US had a consumer based economy. This was probably a diversion from the fact the US manufacturing base has decreased so as to be meaningless. Now they are supporting the buy American slogan. The only problem is there are few American made goods available to buy.
9. In the ancient past disgruntled workers would mount their pitch forks and attack the government. Today's population has access to guns which is more effective than pitch forks. Even today those countries that have severe restrictions against gun ownership the population use rocks and anything that can be used as a weapon.

10. Governments are cognizant of the biblical verse in John 12:8, "for the poor always ye have with you" which creates a huge voter base but is difficult to rectify.
11. After a government's demise it is only a matter of time before the knowledge base dies off or is not accessible. At this time what is known as the dark ages will ensue.

The US democracy is so convoluted that any legislative changes makes it more so. All the listed reason for a demise leads to a military take over. The takeover will be preceded by a period of anarchy. This may not be all bad if the following events occur.

1. All branches of government disbanded.
2. Unions, lawyers, and lobbyists disbanded. Unions if allowed to survive will lose their ability to strike. Lawyers and lobbyist will be restricted in the harm they can engender.
3. Nullify economically punitive laws and departments. Place all government documentation in a historical archive. The knowledge of our demise must be available to help future generations to prevent the same mistakes.
4. Reestablish the house and senate with members who have qualifications to retain existing laws beneficial to the entire society. Many of the laws initiated by special interest groups must be abolished. New laws may require amendments to original constitution.
5. Reestablish the judiciary who will rule on the constitutionally of new laws. The original constitution must not be changed. Amendments may be made to expand or clarify an article to meet modern interpretation.
6. After the country is functioning again under the new order a year divisible by four will be selected to have elections. There will be three political parties.

7. Each state's representative in the transitional government may stand for election. A president will be chosen by popular votes of each state.

8. A means must be devised to replace the members of the Supreme Court. The minor courts should be held accountable for making judicial decisions for monetary gains.

9. State's jury systems should be replaced with a bench of qualified lawyers. A means to measure each member's performance should be implemented.

Reestablish necessary government departments. No department should have the power to harm the economy.

Section-9
UNKNOWN PHENOMIA

MANY OF THE "HOW,WHY,WHEN,WHERE,WHAT" questions are interpreted as messages left by the XTC to enlighten man. If this was truly messages why expend so much effort when the expansion of the Cuneiform and Sanskrit texts would be more effective. This may not have served any purpose since scientists ignore them as they exist today. It was more than likely that the XTC wanted to keep man as deeply in the dark as to their existence as feasible. Is it possible that the XTC has contracted the CIA to help hide their existence ? (joke I hope).

The cuneiform texts are constructed with what could be the oldest instance of an alphabet. The characters of this alphabet are very basic for two reasons. First the characters are easily made with a simple stylus. Second the clay tablets on which they are impressed is a common media. Also the tablets are easy to transport and has an enduring longevity. All the following alphabets were developed by man. Any derived text was inscribed on media that had a lessor longevity than the clay tablets. Most ancient texts were composed by earthly priest that included some information related to the XTC. These texts were at times contaminated with insertions that supported the composer's agenda.

To get back to the HWWWW presentations. Theories exist that purpose that the megalithic structures are messages for man.

This can be thought of the same as the interpretation of pictographs. The interpretation is left to the individual. It is hard to believe that such massive efforts were required when a simple text message would suffice. It is easier to believe the structures were required for occupation and was too expensive to remove when occupation ceased.

The ancient texts of the Indus Valley civilization were composed from oral traditions. They seem to be a non-religious presentation of events that occurred in the distant past. Scientist will claim that the information was changed over time. This only happens in modern times when scientist are involved.

They will stoop so low so as to propose that the information was influenced by TV. (joke)

Scientists main argument against the existence of UFOs is "they can't get here from there". Of course this is based on our feeble attempts of space travel and Einstein's equation's speed limit. In some instances the question is augmented by a second question "where did they come from".

Scientist have computed that the earth, moon, and sun have five points within their plane that are not affected by gravity of either. This says that an object in one of these points would not have to expend any energy to stay in the point. These locations are labeled the Lagrange points. It would be reasonable to assume that other planets could have such points.

For an XTC that has been traversing the universe for millions of years, it would be logical to assume they had plenty of time to get here. Rather than make regular commutes to the home planet, why not install space stations in the Lagrange points. This would have the advantage of not having to worry about planetary catastrophes, and members of the project would have all the comforts of home. There would be no limit to the number of stations or their size. The UFOs would allow travel between the station and earth.

We have not sent any robotic probes to the points and the resolution of telescopes is insufficient to allow monitoring. When we are able to send probes to the points we may stumble onto the existence of the XTC. The population may never know of the event since the government will continue its coverup.

2-Mistrust Of Science

Even as the impediments to scientific thought by religion has ceased, a more insidious impediment has arisen. This impediment takes the form of governmental agencies that deny grants to be awarded to academic scientist that pursue studies in areas not approved by the government. This is many times more harmful than the Spanish Inquisition.

One area of study rejected by the government is that of UFOs. This began following WWII and the ensuing cold war. Reporting of UFOs became so pronounced as to clog the manual switch boards. The intelligence agencies became concerned the UFO reports could clog the switch boards which Russia could use as the time to launch an attack. The CIA was given the task of explaining the cause of the sightings. When it was unsuccessful with an explanation it decided to reduce the number of reported sightings. This was to be achieved with ridicule and threats to those making a UFO report. The air force was tasked with pretending to perform an in depth study of the problem.

The CIA's solution worked well to the detriment of science. Even with the thousands of UFO reports science dare not have an opinion for fear of reprisals. Debunkers are overly employed in the ranks of academic scientist that need their card punched. The government debunking effort is beginning to have the opposite effect as originally intended. The general population is beginning to disbelieve the statements made by the government and science to the point of ridicule. What most people don't realize is that they are only doing what the black project tell them to do. The government can screw up the economy but they can not screw up the black project because they have no control.

In 1947 a crashed UFO was retrieved. The CIA was placed in charge of a governmental cover up. As time passed a black department was created to reverse engineer retrieved UFOs. The CIA still maintains the cover up but it now serves as a contractor to the black department. The black department has become so powerful that even the president does not have knowledge of its operation. If the power source is reverse engineered the countries of the world, including the US, has a lot to worry about. The black department has no allegiance to anyone but itself.

Recent developments in video manipulation has been a much needed tool added to the debunkers toolkit. Now they can easily say that video evidence has been hoaxed. In the past they had to use the unbelievable assertion the video was that of a man in an animal suit or a hub cap thrown into the air. The technology is so good it takes very qualified scientists to determine that its not a hoax. This will result in much information being rejected because it cannot be proven one way or the other. An effective means of dissuading hoaxers is to declare any one convicted will be subject to one year in jail.

Those scientist that are not sub-servant to the government coverup of UFOs participate as advisers in the production of documentary videos and books. They are at a disadvantage since the laws of physics developed by man does not apply. Therefore, they have to develop theories to explain the unknown and unprovable. Of course main stream scientist ridicule the theories even though they do not have an alternate theory. They do not need an explanation because all they need to do is declare speculation.

There have been thousands of books and videos produced on the subject of UFOs. These presentations are well researched. The research is performed by retired scientist and other disciplines that are just as qualified. The academic scientist must not show any interest in fear of grant refusal. Although they are not full fledged debunkers, they should answer questions with "no comment", rather than snide comments that can be interpreted as opposition by their handlers.

The presentations of the unknown will fit very well in the TARRAFORMING EARTH theory. In addition, the modern theories of how life replenished itself after catastrophes can be replaced with theories that are more believable. There will not be any need to stretch Darwin's theory to the breaking point.

The black projects that are involved in reverse engineering crashed UFOs have an advantage because they have the physical hardware. It can be assumed that any discoveries they make will be kept secret.

This can be understood when the information is critical to earthly survival. They definitely cannot allow the politicians who have a low intelligence quotient access to the information. The scary side of the proposition is the black project will be able to subjugate world governments who will become their servants, more so than today. At this time the XTC will become involved because the universal order will be affected. The reasoning behind the tower of Babel may once again come to the fore.

Search for the XTC

The Search for Extra Terrestrial Intelligence (SETI) project has been searching for alien broadcast in the radio spectrum for many years. Since no signals have been detected, the probable reasons are the signals are outside the detection range or the XTC does not use the radio spectrum for communication. If UFOs were using radio for ship to ship communication SETI would have inadvertently detected such transmissions.

Since no ship to ship communications by UFOs have been detected the debunker will interpret the fact to show they do not exist. A similar argument was used following WWII when it was said that UFOs did not exist because they could not be detected by radar. After reversing engineering crashed UFOs earth's scientist developed aircraft that can not be detected by radar.

SETI is a privately funded company that avoids any association with UFOs. In fact, the only time it mentions UFOs is as a debunker, in an effort to protect its project. A possible reason for the avoidance is that the scientists do not wish to tarnish their career in the event the plug is pulled on their funding, and they have to go grant hunting. The scientist has to do nothing but monitor logs for signals that the computer was not programmed to detect. If they keep their mouths shut they don't even have to answer to their peers.

With improved equipment and techniques modern scientist have begun to search for stars with a planetary systems that could include earth like planets. They are careful to say that they don't expect to find intelligent life. Microbial life would be expected and acceptable.

3-Reality Of Ufos

One has to ask himself what event could derail the CIA cover up. One possible event could result when countries that are not controlled by the USA starts to openly investigate the phenomena. Another major event would be, if a UFO landed in a populated area. The possibility of such an event is remotely possible. If the XTC has any reason for direct contact they would have done so in the recent past. The only realistic event that can destroy the cover up is the crash of a UFO in a populated area. It happened in the past, but the CIA quickly erased all evidence. It was then easy to discredit any observers. It would be more difficult to do such today because of the pervasiveness of camera phones. Sure some of the phones could be confiscated, but the law of large numbers comes into play. Those phones not confiscated will have their contents transmitted to the news media in short order. If the news media is not hushed up in time, then millions of people would be aware of the event. Even as the news media is inhibited from dispensing the information, the information will remain in the vast information network to surface at a later date.

The increase in UFO sighting reports indicates the tarraformers have increased their monitoring of the earth. The occupants of the UFOs are reported as grays with a small frame and big eyes. They are reported to have unbelievable mental and physical powers. Some of these powers may be attributed to portable devices. Since they look nothing like a humanoid, they are more than likely not what was reported in ancient texts as XTBs. However, they were encountered by ancient man as is attested to by cave drawings.

There are many movies based on the premise that aliens will someday come to earth with guns blazing in an attempt to gain control. The alien vehicles are seldom depicted similar to what is reported in UFO reports in an effort to provide a distance from reported events. When all earths' weapons proved inadequate, a miracle occurs. The aliens succumb to an earth born virus. The inverse could be possible. The aliens are so efficient they would not resort to physical weapons. They would install a virus and wait a few years for the population to disappear. The XTC has had control of earth for millions of years and will continue to do so in the future.

Many geological events such as earthquakes and volcano eruptions are linked to the observation of UFOs. An erroneous association is that the UFOs caused the event. It is more than likely that they knew it was going to happen and they were there to monitor it. Religion still use the event to claim it is the result of Gods punishment for man's sin.

4-What We Know

Reported characteristics of UFOs are so divergent from the known laws of physics it is human nature to say they are not possible. This leads to an effort to create a believable explanation which very few believe. When the government says that the majority of UFO reports are explained, it is really saying that an explanation has been concocted for the non-believer and a cover for its inaction. The government does not have to supply proof of an explanation. Swamp gas, ball lightning, clouds, airplanes, flares, and Venus are the most common concocted explanations. If none of these work just call it a hoax or hallucination. If a report includes convincing videos or photographs, a "granted" scientist is enrolled as a debunker that states the information is faked.. This selection includes those that have little knowledge of the subject but need their "scientist card" punched.

The UFO phenomena differs from ancient phenomena in that it is here and now. No matter the effort expended by the pseudo scientist of the government to eliminate the problem it does not go away as evidenced by the many current reports.

The following is a list of some of the unexplained characteristics of reported events.

1. They can't get here from there, but they did. Recently more than one significant discovery has been made that

neutrinos can travel faster than the speed of light. If such a discovery proves to be valid then the light speed limit no longer applies which will require many theories to be revisited.

2. Velocity and maneuvers not possible by earthly vehicles due to the law of inertia.
3. Wings and stabilizers never reported.
4. No visual or audio indications of earthly form of propulsion. The instances where a form of propulsion is reported can be attributed to reverse engineering of captured UFOs.
5. Ability to fade into and out of detection by radar and visual. Current stealth technology only minimizes the radar signature and does create invisibility.
6. Ability to function under water.
7. Electromagnetic interference.
8. Mental telepathy independent of language.
9. Ability to erase and create memories.
10. Many more.

It can be seen how most of the listed events can become achievable within a hundred to a thousand years. Even the unbelievable propulsion system may come to pass. At which time we can then traverse the solar system with ease. Another major hurdle is how to eliminate the force of gravity and the force of inertia. Many of the other listed properties may be more easily achieved.

It is more difficult to reconcile the reports of passing through walls. This includes an abductor and his abductee. This suggests inter-dimensional travel which could be imagined for the ufonaut but not his earthly charge. If inter-dimensional travel is ruled out, then one has to ask how does the entity navigate the interior of a building without detection. The easiest explanation is memory creation in the earthly charge and any observer.

It is evident that chemical energy cannot be the power source. The theorized power source is anti-matter of which we know very little since it is not part of our universe. If it were we would observe numerous explosions in the universe when matter and anti-matter come into contact. Even though anti-matter is not coexistent with matter in the universe earth scientist have been able to create miniscule amounts of anti matter at a cost of many millions of dollars per gram. Science has determined that when matter comes into contact anti-matter the result is the release of pure energy. This makes it a possible explanation for the power source of UFOs.

5-OTHER UNKNOWNS

The animal mutilation phenomena are associated with UFOs in some cases. The XTC may be trying to determine the changes man made to the animal population by collecting DNA. Cattle and horses are not indigenous to America but were installed by explorers. The American bison which was installed by the XTC have experienced few mutilations. The instances of human abduction may be attributed to the ETC's collection of DNA to analyze man's evolution or to develop a new man for another planet.

Other unknown phenomena that cause scientists sleepless nights is Bigfoot. The official explanation is that it is a mis-identification of a known animal. Bears are the most used explanation although bears are not indigenous to the area. The scientist violates their creed by saying give us a carcass and we will study it. The reason they feel it is safe to make the statement is that no carcass has ever been found. Maybe what needs to be done is to offer a large bounty and a defense fund against stupidity. A possible consequence of killing a Bigfoot is a UFO encounter while they are also trying retrieve the carcass. Maybe the reason for no carcass is that the XTC installs a new species with a tracking device much more advanced than the crude tracking devices attached to animals by earth's scientist. When the specimen's heart "quits ticking" the XTC will retrieve the carcass for a postmortem or the removal of evidence.

Many non-mainstream scientists to their credit have participated in efforts to develop evidence of Bigfoot's existence all of which have been unsuccessful. A tool used by investigators is the trail camera. If ever successful in capturing an image it will be declared a fake by the debunkers. Since all reports of Bigfoot come about by chance, the required evidence will probably occur when a hunter kills one. This will happen when the hunter is of the type that says, "kill it and lets see what it is". When Bigfoot is proven to exist, the question will arise is he being developed for another planet or as a replacement for man?

Since this collection of unknowns requires more debunkers. One debunker that appeared on an episode of ancient aliens fits the classical definition. His demeanor indicated he desperately needs his scientist card punched. He did not have the unruly hairdo sported by many scientist wannabes, but instead a Mohawk hair cut. He solidified his position about the XTC by making jokes about discussions with his associates over pints of alcohol about various theories. He stated that not once did the discussion include the existence of the XTC.

A recent discovery can be added to the tales of the unknown. A whale carcass was found more than a mile on shore. The question that was asked, but not answered, how did such large specimen appear so far from the sea without signs of transportation. More than likely nothing else will be said about the event.

Repression of information by the government and science is similar to what religion tried to do at the beginning of the scientific method. Religion was able to survive because it continued to support the established belief system. Science on the other hand has deviated from its founding principals. It can only be saved if the government entices the best minds in academia to commence study of unknowns.

The government should leave the endorsement of finding to science, since it is so mistrusted.

Repression of information can last only so long. The advent of UFOs may subvert the repression, but cause other problems. Religion will have to adjust to real entities instead of imaginary entities. Science will have to adjust to the possibility of a more advanced intelligence which is present.

Before we close this section I must share a rumor with you. The XTC determined that the XTBs assigned the task of monitoring earthly TV broadcast were wasting time by playing fantasy football.